Credits

Author – Niraj Choudhari

Disclaims

The contain and the information provide with this book is to be used for educational purpose only. All of the information provided in this book is mean to help and understanding microcontroller and programming to develop electronics hardware with good attitude. The author of this book takes no responsibility for action resulting from the inappropriate usage of learning material content with this book

Inspirer for Engineers

Engineers like to solve there problems. If there are no problem handily available, they will create there own problem.

- Scott Adam

Man is slow, sloppy and brilliant thinker; The machine is fast, accurate and stupid

- William Kelly

Technology…the knack of so arranging the world that we don't have to experience it.

- Max Frisch

Technology presumes there's just one right way to do things and there never is.

- Robert M. Pirsing

Engineers are lazy, still they discover the electrons in brain

- Niraj Choudhari

Crazy Circuits

Practices for beginners

Niraj D. Choudhari

Volume1

Special Thanks

To my mother and father

To Students

-Niraj Choudhari

Thanks To:

- My dear students, to supporting my 1st book on technical circuits.

- God, who looks after me.

- Rushabh Kidile, Yash Sawarkar, Ashwin Bhajan, Gunjan Raut,, have always inspired me for writing this book and there support.

- Palash Dalal, Pritish Kumbhare, Pranay Jaiswal, who are the first reader and advisor of this book.

- Shruti Sangidwar and Parv Choubey was as backbone support at all time for completion of this book.

- Shyam Prasad, Neeraj Yadav, Ranjit Kumar, Dilip Pardhi, Gajendra Patle and Amol Shirbhate have always supported me technically and also teach me develop me as technical person.

- The amazing friends and my family members, for developing my interest to write the book.

- My mother Pushpa and my father Dnyaneshwar for being in my life and always support me by tell "Baccha, don't be tense if it's not getting complete, Nothing to loose" when i fees low.

-Niraj Choudhari

Index

Serial No.	Name	Page No.
1	LED blinker	8
2	Multi junction LED blinker	11
3	Seven Segment Display	14
4	Push to ON-OFF Switch	19
5	Push to ON-OFF Switch	23
6	Introducing to ADC	28
7	Power regulator display	33
8	Relay interfacing	40
9	DC motor interfacing	45
10	Propeller display	50
11	Fire detector	56
12	Day light detector	61
13	Auto day shade light	66
14	Obstacle detector	70
15	Introduction to PWM	76
16	LED Controlled using PWM via Button	80
17	LCD Brightness controlling	84
18	DC motor speed controlling using PWM	89
19	Voltage Step-down Without	94

	Transformer	
20	Laser security	97
21	Temperature sensor interfacing	101
22	PIR Sensor	106
23	4x4x4 LED cube	110
24	Sound Sensor Interfacing	126
25	Accelerometer Interfacing	131
26	Appendix	137
27	Header file	156

Experiment - 1

Led Blinker

Led blinker is simple circuit to introduce interest in digital electronics. Its basic circuit comprise of LED and AVR. There are two types to make led blinker they are time linear and time non-liner, both the type based on delay and delay is important factor in LED blinker. In time liner blinker delay is predefine and it is constant but in time non-liner the delay is change accordance with time its is random blinking of LED.

The blinking can be perform on variety of controller but we are working on ATmega8, ATmega16 and Atmega32 according to program flash required. In this circuit we are using ATmega16 and dumping simple program of LED blinker. The blinking is nothing but simply on – off the LED using delay. In this program port- A is configured to connect with LED with common ground configuration when port-A is set to 5 volt LED will ON and in same state for delay and when port-A is set to 0 LED will set to OFF till delay and same will repeat because it is in repeat loop and we well get blinking effect.

The line mention " PORTA= 0b11111111; " state that we are assigning logic 5 volte to port A and every pin in port A from PORTA0 – PORTA7 is high. We can create different pattern of blinking by making changes in PIN out for ex. " PORTA= 0b01010101; " , " PORTA= 0b11110000; "

The line mention "DDRA = 0b11111111" This stands for Directive register to be select; the port we are using is output port or input port, where 1 stands for output and 0 stands for input.

Programming is done in "AVRstudio4 " and header are mention in program are mention at the end. The component are used are given below.

Component details

1) AVR ATmega16 IC3
2) 7805 IC2
3) LED
4) Crystal 16 MHz
5) 9 volt battery cap
6) ISP burner

Program

(Time linear)

```c
#include<avr/io.h>
#include<util/delay.h>
main()
{
DDRA = 0b11111111;
PORTA= 0b00000000;
    while(1)
    {
        PORTA= 0b11111111;        " line – a "
        _delay_ms(2000);
        PORTA= 0b00000000;        "line – b"
        _delay_ms(2000);
    }

}
```

Schismatic Diagram

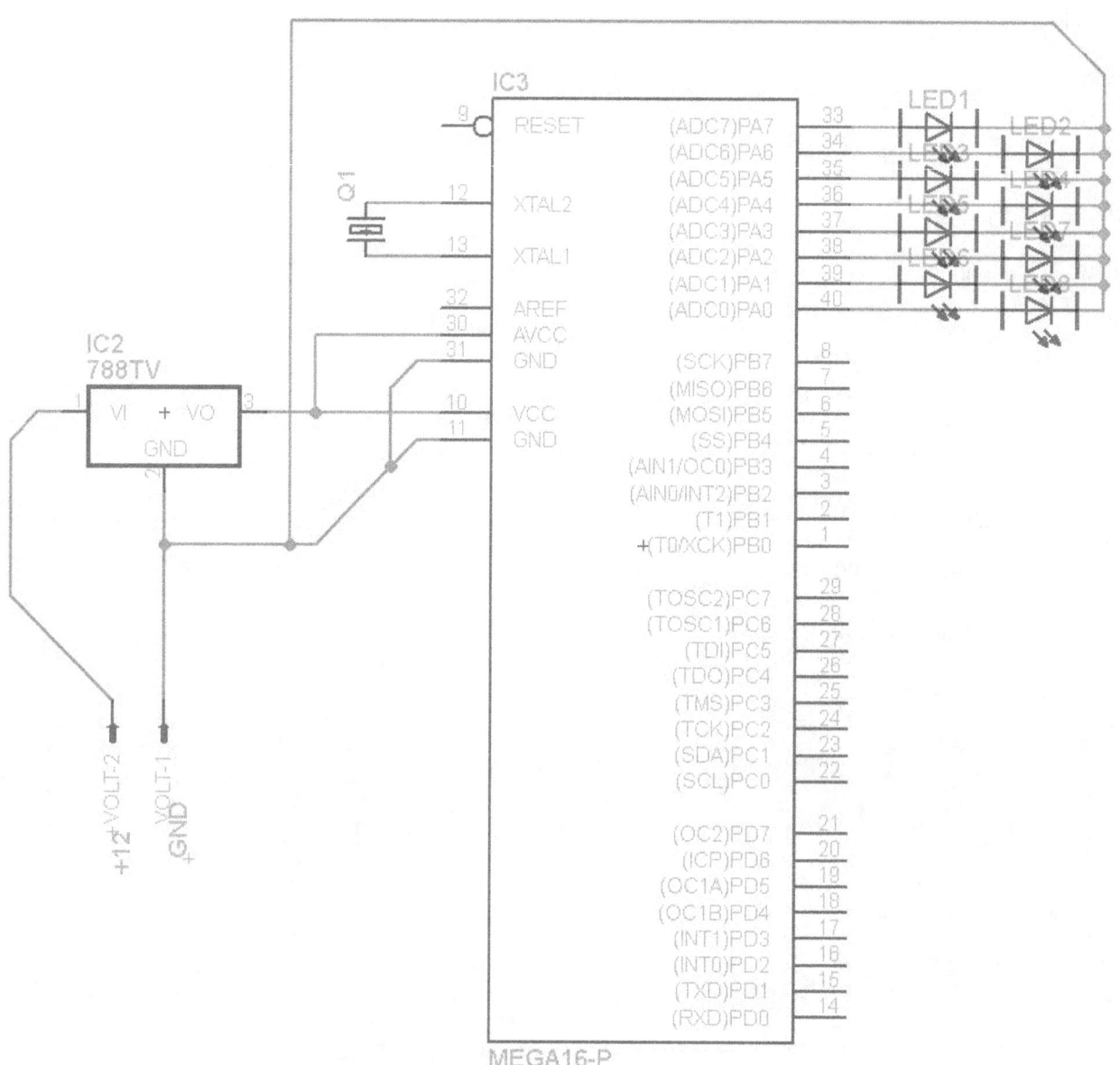

Experiment - 2

Multi junction led blinking

The basic circuit we worked is on led of single junction; there are varieties of LED available in market single junction multi junction. Multi junction LED comes for dual color two color or three color to interface them is same as above circuit but we are going to use different ports. This circuit is also time linear because its working on delay.

The blinking can be perform on variety of controller but we are working on ATmega8, ATmega16 and Atmega32 according to program flash required. In this circuit we are using ATmega16 and dumping simple program of LED blinker. The blinking is nothing but simply on – off the LED using delay. In this program port- A and port-B is configured to connect with LED with common ground configuration when port-A is set to 5 volt LED will ON and in same state for delay and when port-A is set to 0 LED will set to OFF till delay and again Port–B will set to 5 volt to ON led but for this time for different color and will on till delay and get off, here you well get the effect of dual color in same pattern and same will repeat because it is in repeat loop and we well get dual color blinking effect.

The line mention " PORTA= 0b11111111; " & " PORTB= 0b11111111; state that we are assigning logic 5 volte to port A & portB and every pin in port A & B from PORTA0, B0 – PORTA7, B7 is high. We can create different pattern of blinking by making changes in PIN out for ex. " PORTA= 0b01010101; " . " PORTB= 0b10101010; " , " PORTA= 0b11110000; " " PORTB= 0b00001111; "

The line mention "DDRA = 0b11111111" & "DDRB = 0b11111111" This stands for Directive register to be select; the port we are using is output port or input port, where 1 stands for output and 0 stands for input.

Programming is done in "AVRstudio4 " and header are mention in program are mention at the end. The component are used are given below.

Component details

1) AVR ATmega16 IC3
2) 7805 IC2
3) Dual junction LED
4) Crystal 16 MHz
5) 9 volt battery cap

6) ISP Burne

Program

(Time linear)

#include<avr/io.h>

#include<util/delay.h>

main()

{

DDRA = 0b11111111;

DDRB = 0b11111111;

PORTA= 0b00000000;

PORTB= 0b00000000;

    while(1)

    {

        PORTA= 0b11111111;   // port set to logical 5 volt " line – a "

        _delay_ms(2000);   // delay

        PORTA= 0b00000000;   // port set to logical 0 volt "line – b"

        _delay_ms(2000);   // delay

        PORTB= 0b11111111;   // port set to logical 5 volt " line – c "

        _delay_ms(2000);   // delay

        PORTB= 0b00000000;   // port set to logical 0 volt "line – d"

    }

}

Schismatic Diagram

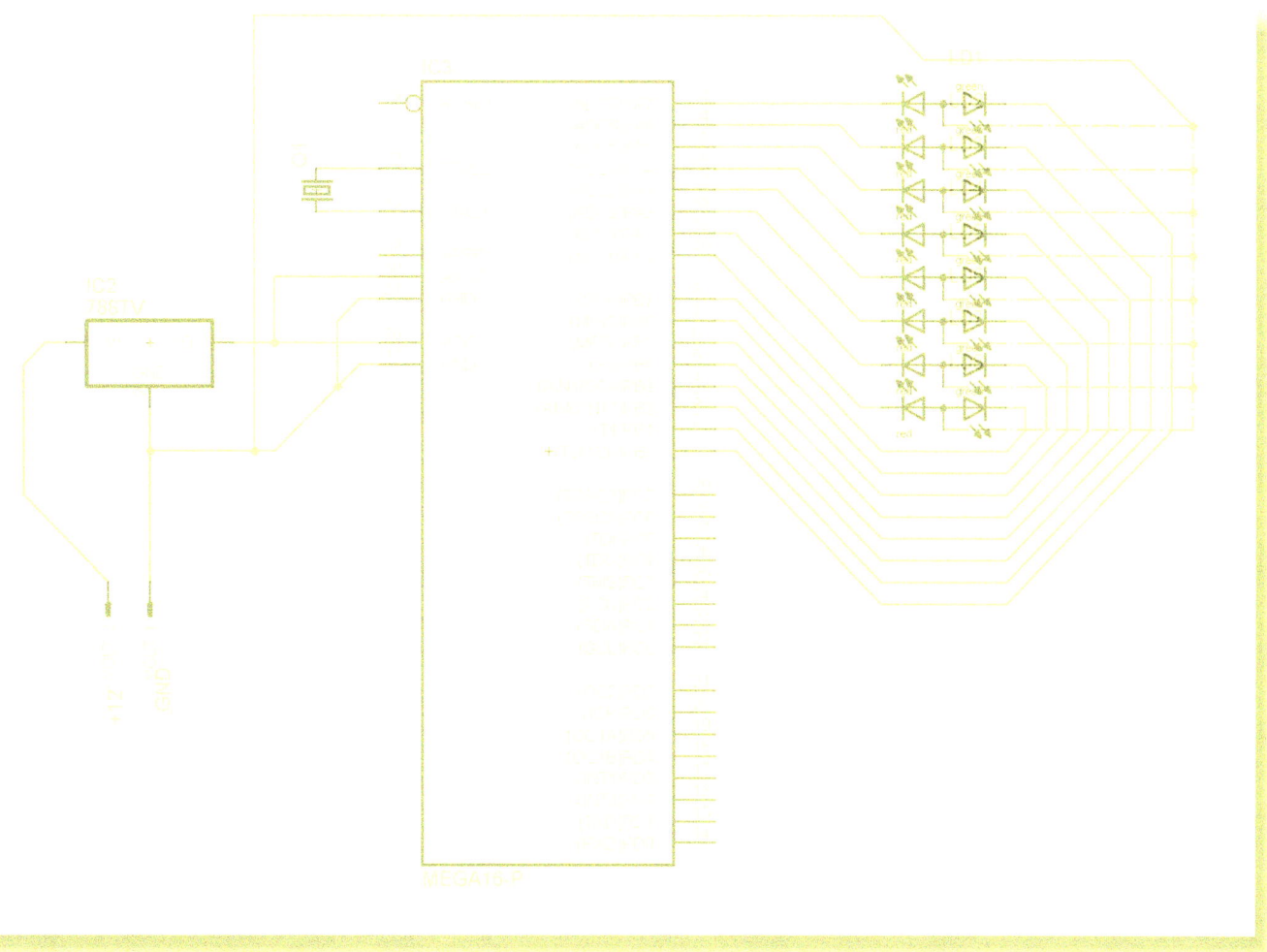

Experiment – 3

SEVEN Segment Display

The *7-segment display*, also written as "seven segment display", consists of seven LEDs (hence its name) arranged in a rectangular fashion as shown. Each of the seven LEDs is called a segment because when illuminated the segment forms part of a numerical digit (both Decimal and Hex) to be displayed. An additional 8th LED is sometimes used within the same package thus allowing the indication of a decimal point, (DP) when two or more 7-segment displays are connected together to display numbers greater than ten, but we are practicing on 1 7-segnment display.

Each one of the seven LEDs in the display is given a positional segment with one of its connection pins being brought straight out of the rectangular plastic package. These individually LED pins are labeled from A through to G representing each individual LED. The other LED pins are connected together and wired to form a common pin. So by forward biasing the appropriate pins of the LED segments in a particular order, some segments will be light and others will be dark allowing the desired character pattern of the number to be generated on the display. This then allows us to display each of the ten decimal digits 0 through to 9 on the same 7-segment display.

The displays common pin is generally used to identify which type of 7-segment display it is. As each LED has two connecting pins, one called the "Anode" and the other called the "Cathode", there are therefore two types of LED 7-segment display called: **Common Cathode** (CC) and **Common**

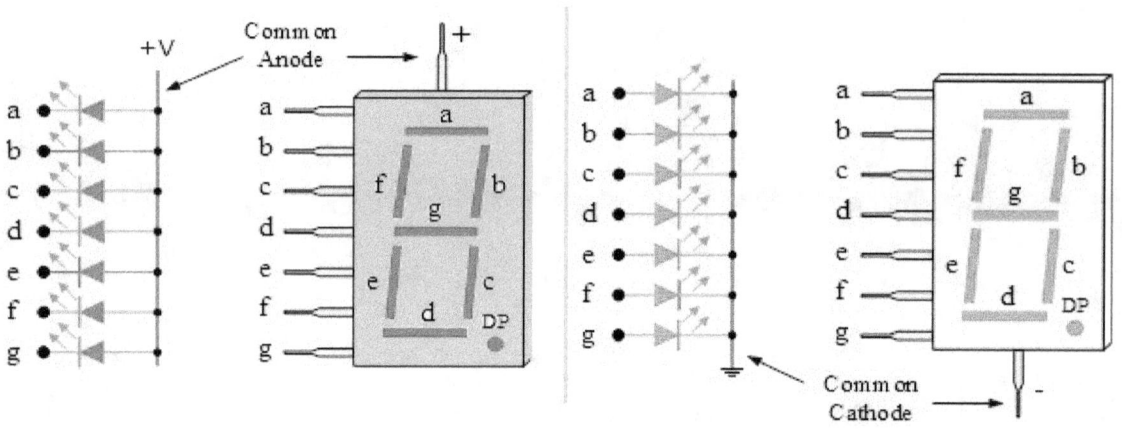

Anode (CA).

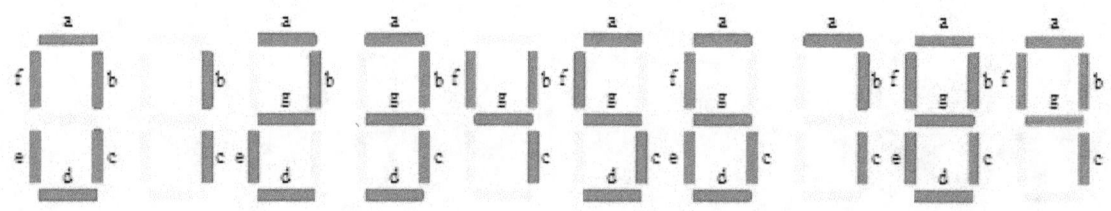

This seven segment is interface with the AVR atmega16 as we are requiring less flash memory. Seven segment is interface on PORTA of AVR, the patterns we are generating to ON – OFF led are in sequence to generate the numeric number from 0 – 9. Interfacing is done direct with AVR without any basing register because current output capacity of AVR is comparatively low.

The configuration is done as shown respectively.

The line mention " PORTA = 0b11111111; " state that we are assigning logic 5 volte to port A and every pin in port A from PORTA0 – PORTA7 is high.

The line mention "   " is use to clear the latching ON LED to OFF state otherwise it will show false numeric value

Connection are mention as below to make changes in your program, the output format is mention in yellow highlight you have to replace "x" with 1 or 0 to change the state of LED to ON-OFF.

0b	x	x	x	x	x	x	x	x	<- PORT OUTPUT FORMAT
	0	1	2	3	4	5	6	7	<- PIN OF CONTROLLER TO OPERATE
	DP	G	F	E	D	C	B	A	<- 7SEGMENT PINS

The line mention "DDRA = 0b11111111" This stands for Directive register to be select; the port we are using is output port or input port, where 1 stands for output and 0 stands for input.

Programming is done in "AVRstudio4" and header are mention in program are mention at the end. The component are used are given below.

Component details

1) AVR ATmega16 IC3
2) 7805 IC2
3) Seven Segment display
4) Crystal 16 MHz
5) 9 volt battery cap
6) ISP Burner

Program

```c
#include<avr/io.h>
#include<util/delay.h>
main()
{
DDRA = 0b11111111;
PORTA= 0b00000000;
    while(1)
    {
        PORTA= 0b00000000;   // port set to logical 5 volt " Null "
        _delay_ms(2000);     // delay
        PORTA= 0b00111111;   // port set to logical 0 volt "0"
        _delay_ms(2000);     // delay
        PORTA= 0b00000000;   // clear display
        PORTA= 0b00000110;   // port set to logical 5 volt " 1 "
        _delay_ms(2000);     // delay
        PORTA= 0b00000000;   // clear display
        PORTA= 0b01011011;   // port set to logical 0 volt "2"
        _delay_ms(2000);     // delay
        PORTA= 0b00000000;   // clear display
        PORTA= 0b01001111;   // port set to logical 0 volt "3"
        _delay_ms(2000);     // delay
        PORTA= 0b00000000;   // clear display
        PORTA= 0b01100110;   // port set to logical 0 volt "4"
        _delay_ms(2000);     // delay
        PORTA= 0b00000000;   // clear display


        PORTA= 0b01101101;   // port set to logical 0 volt "5"
```

```
        _delay_ms(2000);
        PORTA= 0b00000000;
        PORTA= 0b01111101;                    "6"
        _delay_ms(2000);
        PORTA= 0b00000000;
        PORTA= 0b00000111;                    "7"
        _delay_ms(2000);
        PORTA= 0b00000000;
        PORTA= 0b11111111;                    "8"
        _delay_ms(2000);
        PORTA= 0b00000000;
        PORTA= 0b01100111;                    "9"
    }

}
```

Schismatic Diagram

Experiment – 4

Push to ON-OFF Switch

The simple and good learning for beginners in the field of AVR programming. I hope that you already try first experiment of Blinking LED using Atmega16 and Atmel Studio. In most of the embedded electronic practicals you may want to use a push button switch to give user inputs to the microcontroller. To use a push button switch with a microcontroller, first you should configure the corresponding pin as input. Then we can easily read the status of that input pin and make required decisions. There are two types of push button switches Push To On and Push To Off, here we are using Push To On switch. In this tutorial a press at the switch turns ON the LED and second switch to turn off the LED.

A push switch is a momentary or non-latching switch which causes a temporary change in the state of an electrical circuit only while the switch is physically actuated. An automatic mechanism (i.e. a spring) returns the switch to its default position immediately afterwards, restoring the initial circuit condition. A push to make switch allows electricity to flow between its two contacts when held in. When the button is released, the circuit is broken. This type of switch is also known as a Normally Open (NO) Switch. (Examples: doorbell, computer case power switch, calculator buttons, individual keys on a keyboard)

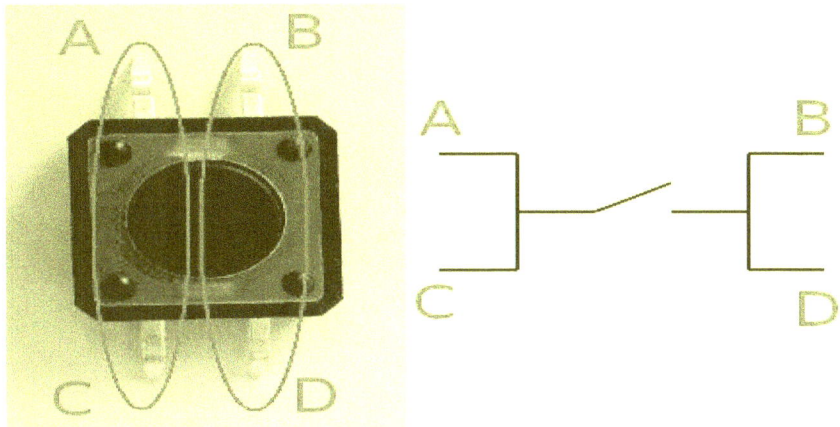

The LED can be ON – OFF in two conditions of switching they are single switch to on–off and dual switch to on-off. We are using dual switch to change state ON to OFF and OFF to ON.

Delay is important factor to detect the switch, improper delay or avoid of delay can give the false statement to the practical output. In this experiment we use two different ports for LED and Switch. We can use same port to switch and LED it just to make simple change in program.

In this experiment we use ATmega16 due to low flash requirement. In this we use PORT-A as input port which is connected to switch and PORT-B as output port which is connected to LED. We are giving logical 5 volt to

PORTB to glow LED and we are giving logical 5 volt to input PORT-A via switch to detect the switching action to microcontroller.

When you push the 1st button it will trigger the 5 volt to the IC ATmega16 to the PORTA to sense input detection from the 1st button to turn ON LED and PORTB0 pin will get high and LED get glow which is connected on PORTB. When 2nd button is trigger same 5 volt gets detect by ATmega16 which turn OFF the LED.

The line mention " PORTA= 0b11111111; " & " PORTB= 0b00000000; state that we are assigning logic 5 volte and logic 0 volt to port A & port B respectively and every pin in port A & B from PORTA0, B0 – PORTA7, B7 is high and low respectively. We can assign any pin of PORTA to turn on and off the LED.

The line mention "DDRA = 0b11111111" & "DDRB = 0b00000000" This stands for Directive register to be select; the port we are using is output port or input port, where 1 stands for output and 0 stands for input.

Programming is done in "AVRstudio4 " and header are mention in program are mention at the end. The component are used are given below.

WARNING

DO NOT GIVE MORE THAN 5 VOLT @500 Milliamp TO SWITCH, and volt must be greater than 2.85 volt otherwise controller will wont sense the sensitivity of switch triggering

Component details

1) AVR ATmega16 IC3
2) 7805 IC2
3) Push to ON switch
4) LED
5) Crystal 16 MHz
6) 9 volt battery cap
7) ISP Burner

Program

```c
#include<avr/io.h>
#include<util/delay.h>
main()
{
DDRA = 0b00000000;
DDRB = 0b11111111;
PORTA= 0b1111111;
PORTB = 0b0000000;
    while(1)
    {
        if(PINA == 0b00000001)
        {
            PORTB = 0b00000001;
            _delay_ms(150);
        }
        if(PINA == 0b00000010)
        {
            PORTB = 0b00000000;
            _delay_ms(150);
        }
    }
}
```

Schismatic Diagram

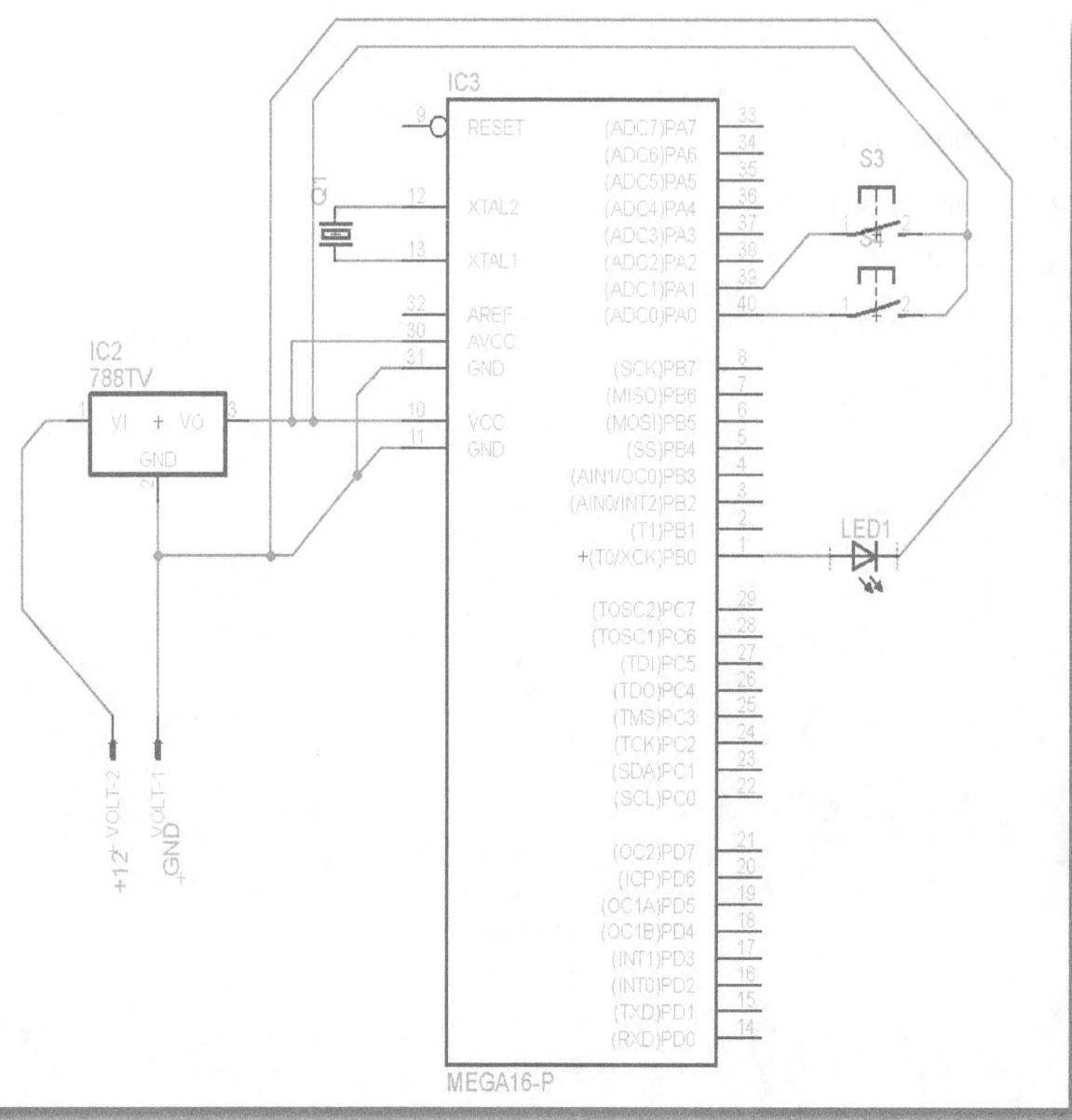

Experiment – 5

16x2 LCD (Liquid Crystal Display) Interfacing

LCD (Liquid Crystal Display) screen is an electronic display module and find a wide range of applications. A 16x2 LCD display is very basic module and is very commonly used in various devices and circuits. These modules are preferred over seven segments and other multi segment LEDs.

A **16x2 LCD** means it can display 16 characters per line and there are 2 such lines. In this LCD each character is displayed in 5x7 pixel matrix. This LCD has two registers, namely, Command and Data.

The command register stores the command instructions given to the LCD. A command is an instruction given to LCD to do a predefined task like initializing it, clearing its screen, setting the cursor position, controlling display etc. The data register stores the data to be displayed on the LCD. The data is the ASCII value of the character to be displayed on the LCD. Click to learn more about internal structure of a LCD.

It consists of 16 rows and 2 columns of 5×7 or 5×8 LCD dot matrices. The module were are talking about here is type number JHD162A which is a very popular one . It is available in a 16 pin package with back light ,contrast adjustment function and each dot matrix has 5×8 dot resolution

Pin Detail

VEE pin is meant for adjusting the contrast of the LCD display and the contrast can be adjusted by varying the voltage at this pin. This is done by connecting one end of a POT to the Vcc (5V), other end to the Ground and connecting the center terminal (wiper) of of the POT to the VEE pin. See the circuit diagram for better understanding.

The JHD162A has two built in registers namely data register and command register. Data register is for placing the data to be displayed , and the command register is to place the commands. The 16×2 LCD module has a set of commands each meant for doing a particular job with the display. We will discuss in detail about the commands later. High logic at the RS pin will select the data register and Low logic at the RS pin will select the command register. If we make the RS pin high and the put a data in the 8 bit data line (DB0 to DB7) , the LCD module will recognize it as a data to be displayed . If we make RS pin low and put a data on the data line, the module will recognize it as a command.

R/W pin is meant for selecting between read and write modes. High level at this pin enables read mode and low level at this pin enables write mode.

EN pin is for enabling the module. A high to low transition at this pin will enable the module.

DB0 to DB7 are the data pins. The data to be displayed and the command instructions are placed on these pins.

LED+ is the anode of the back light LED and this pin must be connected to Vcc through a suitable series current limiting resistor. LED- is the cathode of the back light LED and this pin must be connected to ground.

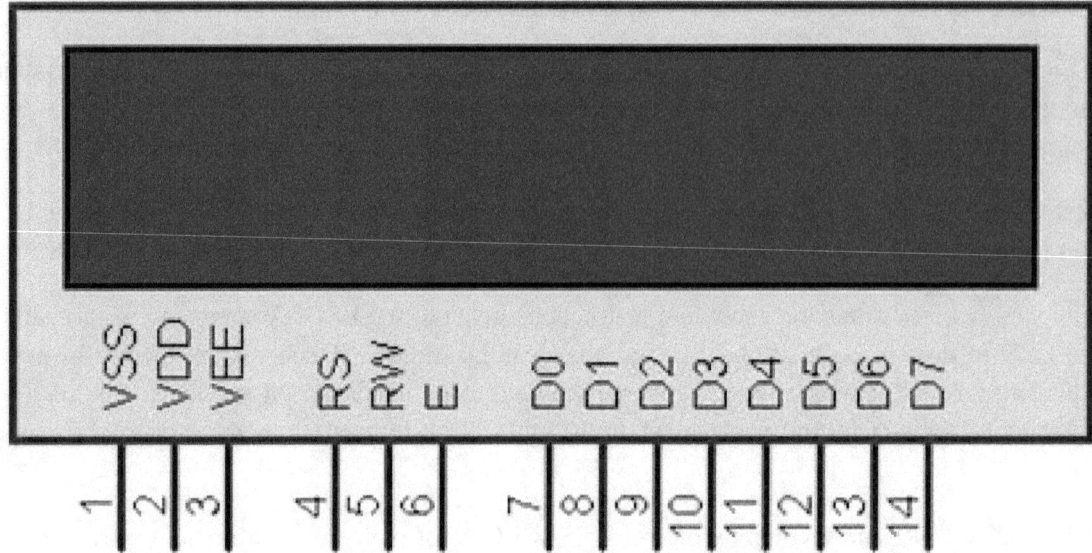

In this experiment we are interfacing LCD with ATmega16 as we are programming to print the statement "Hello". There are many was to print in LCD using commend like curser blinking left to right etc. we can also print numeric value in LCD

To support the LCD program we use header file name "Swits.H" which is mention at the end, including of this header is essential without this header it program won't work.

Pin **R/W** is grounded because we are only practicing on LCD writing not on LCD reading.

Command mention "lcd_init()" is important command to initialize the LCD command from header file.

Command mentiion "lcd_clrscr()" use to clear the content of the LCD.

Command mention "lcd_prints("xxxxxxx")" is use to print the contain which is in the double inverted coma but it has three different data types they are.

1) "Prints " is use for string data type.

2) "Printi" is use for integer data type.

3) "Printc" is use for character data type.

Command mention "lcd_goto(0,1)" is use to move the cursor at mention axis in X,Y coordinates such as lcd_goto(x,y).

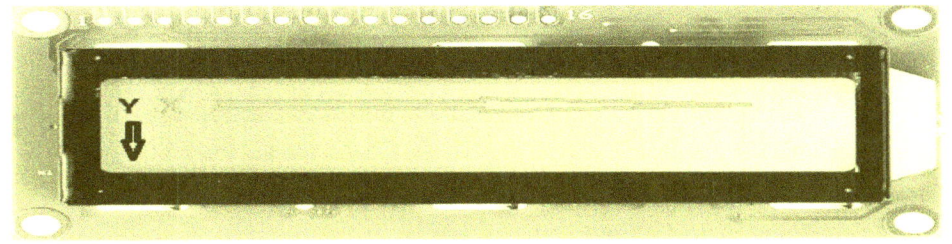

X axis is use to indicate the row and Y axis is use to indicate the Colum.

Programming is done in "AVRstudio4 " and header are mention in program are mention at the end. The component are used are given below.

WARNING

The header we used having predefine PORT to LCD pin connection on PORT – B as mention below, Inter changing of connection can create malfunctioning of LCD or may not run

PORT- B	LCD Pin
PB0 --	RS
PB1 --	EN
PB4 --	D4
PB5 --	D5
PB6 --	D6
PB7 --	D7

Component details

1) AVR ATmega16 IC3
2) 7805 IC2
3) 16x2 LCD
4) LED
5) Crystal 16 MHz
6) 12 DC adptor
7) 10K Preset

8) ISP burner

Program

```c
#include<avr/io.h>
#include<lcd/Swits.h>
#include<util/delay.h>
main()
{
    lcd_init();        // this command is use to initialize of LCD
    while(1)
    {
        lcd_clrscr();
        lcd_goto(0,1);
        lcd_prints("****Hello****");
        lcd_goto(0,2);
        lcd_prints("Niraj");
        _delay_ms(250);
    }
}
```

Schismatic Diagram

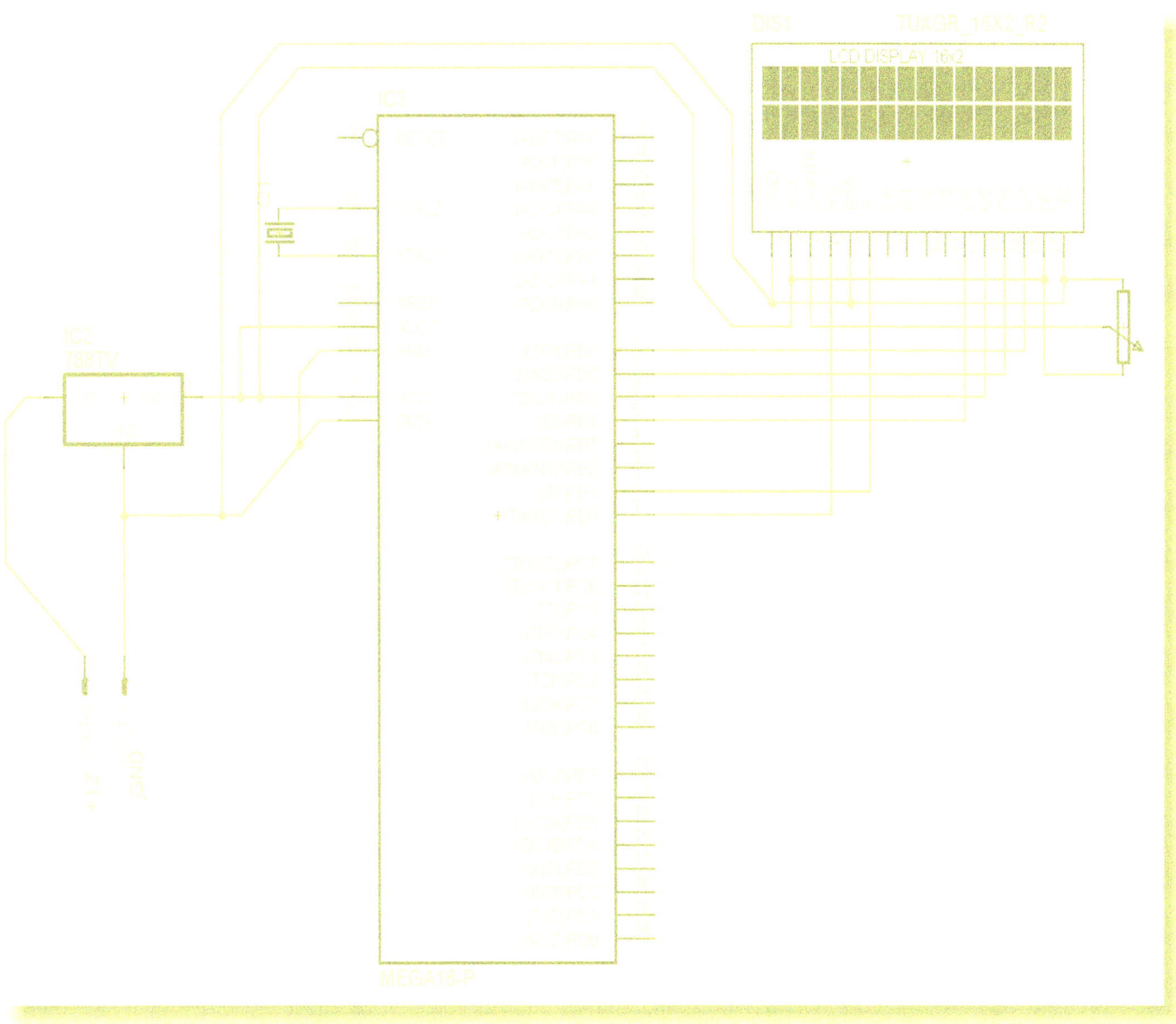

Experiment – 6

Introduction to ADC

Most real world data is analog. Whether it be temperature, pressure, voltage, etc, their variation is always analog in nature. For example, the temperature inside a boiler is around 800°C. During its light-up, the temperature never approaches directly to 800°C. If the ambient temperature is 400°C, it will start increasing gradually to 450°C, 500°C and thus reaches 800°C over a period of time. This is an analog data. Now, we must process the data that we have received. But analog signal processing is quite inefficient in terms of accuracy, speed and desired output. Hence, we convert them to digital form using an Analog to Digital Converter (ADC).Interfacing Sensors In general, sensors provide with analog output, but a MCU is a digital one. Hence we need to use ADC. For simple circuits, comparator op-amps can be used. But even this won't be required if we use a MCU. We can straightaway use the inbuilt ADC of the MCU. In ATMEGA16/32, PORTA contains the ADC pins. The AVR features inbuilt ADC in almost all its MCU. In ATMEGA16/32, PORTA contains the ADC pins.

The channel implies that there are 8 ADC pins are multiplexed together. You can easily see that these pins are located across PORTA (PA0…PA7).10 bit resolution implies that there are $2^{10} = 1024$ steps

Here we have to give reference voltage to pin Aref, there are 2 type of reference we can apply 2.5 volt either 5volt.

Suppose we use a 5V reference. In this case, any analog value in between 0 and 5V is converted into its equivalent ADC value. The 0-5V range is divided into $2^{10} = 1024$ steps. Thus, a 0V input will give an ADC output of 0, 5V input will give an ADC output of 1023, whereas a 2.5V input will give an ADC output of around 512. This is the basic concept of ADC.

The ADC of the AVR converts analog signal into digital signal at some regular interval. This interval is determined by the clock frequency. In general, the ADC operates within a frequency range of 50kHz to 200kHz. But the CPU clock frequency is much higher (in the order of MHz). So to achieve it, frequency division must take place. The prescaler acts as this division factor. It produces desired frequency from the external higher frequency. There are some predefined division factors – 2, 4, 8, 16, 32, 64, and 128. For example, a prescaler of 64 implies F_ADC = F_CPU/64. For F_CPU = 16MHz, F_ADC = 16M/64 = 250kHz.

Now, which frequency to select? Out of the 50kHz-200kHz range of frequencies, which one do we need?. There is a trade-off between frequency and accuracy. Greater the frequency, lesser the accuracy and vice-versa. So, if your application is not sophisticated and doesn't require much accuracy, you could go for higher frequencies.

The ADC needs a reference voltage to work upon. For this we have a three pins AREF, AVCC and GND. We can supply our own reference voltage across AREF and GND. For this, **choose the first option**. Apart from this case, you can either connect a capacitor across AREF pin and ground it to prevent from noise, or you may choose to leave it unconnected. If you want to use the VCC (+5V), **choose the second option**. Or else, **choose the last option** for internal Vref.

Detail

The ADC converts an analog input voltage to a 10-bit digital value through successive approximation. The minimum value represents GND and the maximum value represents the voltage on the AREF pin minus 1 LSB. Optionally, AVCC or an internal 2.56V reference voltage may be connected to the AREF pin by writing to the REFSn bits in the ADMUX Register.

The internal voltage reference may thus be decoupled by an external capacitor at the AREF pin to improve noise immunity. The analog input channel and differential gain are selected by writing to the MUX bits in ADMUX. Any of the ADC input pins, as well as GND and a fixed bandgap voltage reference, can be selected as single ended inputs to the ADC.

A selection of ADC input pins can be selected as positive and negative inputs to the differential gain amplifier. If differential channels are selected, the differential gain stage amplifies the voltage difference between the selected input channel pair by the selected gain factor. This amplified value then ADC CONVERSION COMPLETE IRQ 8-BIT DATA BUS 15 0 ADC MULTIPLEXER SELECT (ADMUX) ADC CTRL. & STATUS REGISTER (ADCSRA) ADC DATA REGISTER (ADCH/ADCL) MUX2 ADIE ADEN ADSC ADATE ADIF ADIF MUX1 MUX0 ADPS2 ADPS1 ADPS0 MUX3 CONVERSION LOGIC 10-BIT DAC + - SAMPLE & HOLD COMPARATOR INTERNAL 2.56V REFERENCE MUX DECODER MUX4 AVCC ADC7 ADC6 ADC5 ADC4 ADC3 ADC2 ADC1 ADC0 REFS1 REFS0 ADLAR + - CHANNEL SELECTION GAIN SELECTION ADC[9:0] ADC MULTIPLEXER OUTPUT GAIN AMPLIFIER AREF BANDGAP REFERENCE PRESCALER SINGLE ENDED / DIFFERENTIAL SELECTION GND POS. INPUT MUX NEG. INPUT MUX TRIGGER SELECT ADTS[2:0] INTERRUPT FLAGS START 206 2466T–AVR–07/10 ATmega16(L) becomes the analog input to the ADC.

If single ended channels are used, the gain amplifier is bypassed altogether. The ADC is enabled by setting the ADC Enable bit, ADEN in ADCSRA. Voltage reference and input channel selections will not go into effect until ADEN is set. The ADC does not consume power when ADEN is cleared, so it is recommended to switch off the ADC before entering power saving sleep modes.

The ADC generates a 10-bit result which is presented in the ADC Data Registers, ADCH and ADCL. By default, the result is presented right adjusted, but can optionally be presented left adjusted by setting the ADLAR bit in ADMUX. If the result is left adjusted and no more than 8-bit precision is required, it is sufficient to read ADCH. Otherwise, ADCL must be read first, then ADCH, to ensure that the content of the Data Registers belongs to the same conversion. Once ADCL is read, ADC access to Data Registers is blocked.

This means that if ADCL has been read, and a conversion completes before ADCH is read, neither register is updated and the result from the conversion is lost. When ADCH is read, ADC access to the ADCH and ADCL Registers is re-enabled. The ADC has its own interrupt which can be triggered when a conversion completes. When ADC access to the Data Registers is prohibited between reading of ADCH and ADCL, the interrupt will trigger even if the result is lost.

Practicing ADC

We are use to connect 10k pot to the analog port of ATmega16. The analog variation will display in LCD. To simplify the ADC program we make header "swits" instead of calling register making complicate program.

The header file mention contain all predefine code for the ADC we just initialize the ADC and call the function from header.

Command mention "lcd_init()" is important command to initialize the LCD command from header file.

Command mention "init_adc()"()" is important command to initialize the ADC command from header file.

Command mention "x=read_adc(0)" this command reading the ADC value and assigning to 'x' whose data type is integer.

Programming is done in "AVRstudio4" and header are mention in program are mention at the end. The component are used are given below.

WARNING

Do not give supply more than 5 volt to the ADC port of ATmega16

Component details

1) AVR ATmega16 IC3
2) 7805 IC2
3) 10k Pot
4) 10K preset
5) LCD
6) Crystal 16 MHz
7) 9 volt battery cap
8) ISP Burner

Program

#include<avr/io.h>

#include<lcd/Swits.h>

#include<util/delay.h>

main()
{
    lcd_init();
init_adc();
int x;
while(1)
{
    x=read_adc(0);
    lcd_clrscr();
    lcd_goto(0,1);
    lcd_prints("2.5 volt");
    _delay_ms(150);

}
}

Schismatic Diagram

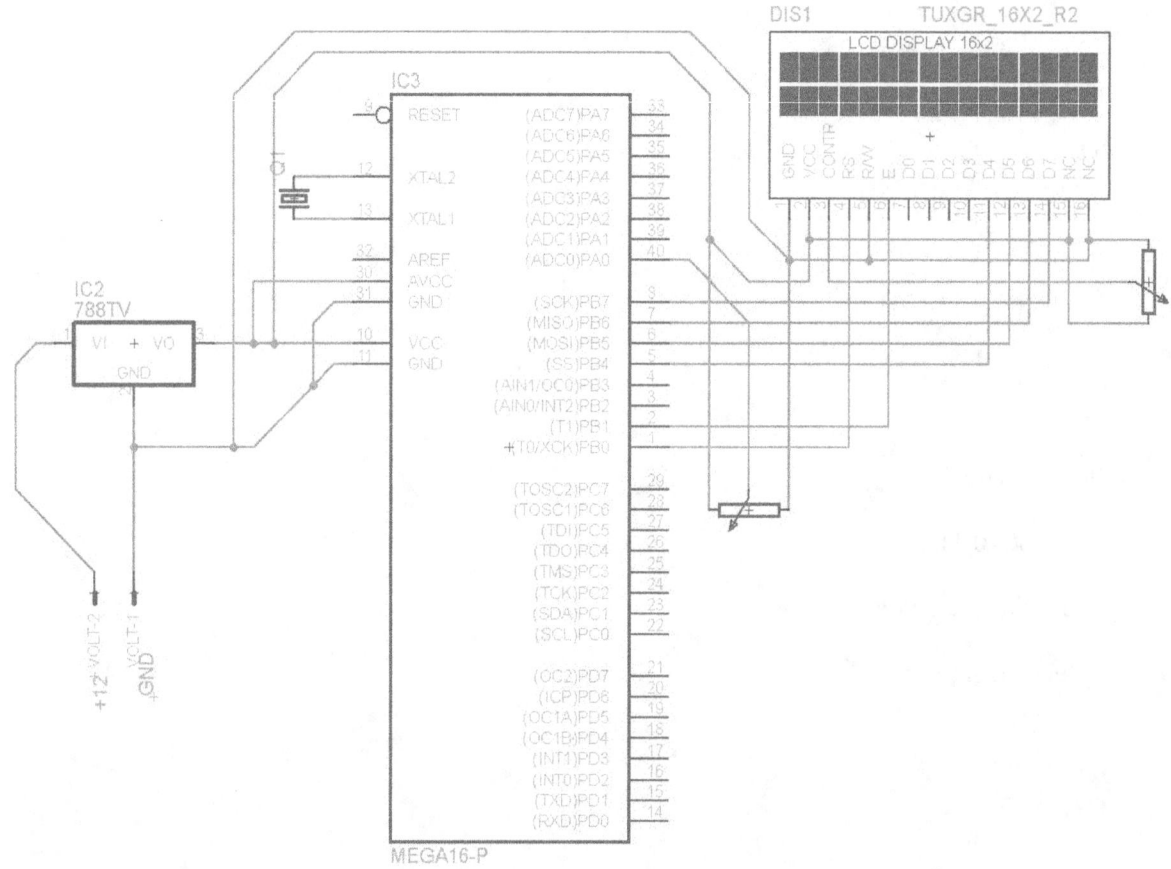

Experiment – 7

Power regulator display

In most of our electronic projects we need a power supply for converting mains AC voltage to a regulated DC voltage. For making a power supply designing of each and every component is essential. The digital circuits work on logical 5 volts, to test the digital circuits we require the supply source of 5volts @ 0.5 amp.

There are verities of IC available to regulate the power supply coming from step-down transformer. For positive regulation 78XX series is available to regulate the voltage at mention output. There is also negative voltage regulator that is 79XX series required for Operational amplifiers but both the series are used for fixed regulation, for variable voltage regulation LM371 IC is available.

The **78xx** is a family of self-contained fixed linear voltage regulator integrated circuits. The 78xx family is commonly used in electronic circuits requiring a regulated power supply due to their ease-of-use and low cost. For ICs within the family, the **xx** is replaced with two digits, indicating the output voltage (for example, the 7805 has a 5-volt output, while the 7812 produces 12 volts). The 78xx line are positive voltage regulators they produce a voltage that is positive relative to a common ground. 78xx series ICs do not require additional components to provide a constant, regulated source of power, making them easy to use, as well as economical and efficient uses of space. Other voltage regulators may require additional components to set the output voltage level, or to assist in the regulation process. 78xx series ICs have built-in protection against a circuit drawing too much current. They have protection against overheating and short-circuits, making them quite robust in most applications. In some cases, the current-limiting features of the 78xx devices can provide protection not only for the 78xx itself, but also for other parts of the circuit.

78XX	Output Voltage	Input Voltage
7805	+5	7.3
7806	+6	8.3
7808	+8	10.5
7809	+9	12.5
7812	+12	14.6
7815	+15	17.7
7818	+18	21.0
7824	+24	27.1

The 79xx voltage regulators are very commonly used in electronic circuits. The main purpose of this IC is to supply required regulated negative voltage to the circuits. The **xx** is replaced with two digits, indicating the output voltage (for example, the 7905 has a -5volt output, while the 7912 produces -12 volts). IC 79xx can supply a constant negative voltage output, in spite of any voltage fluctuations in its input voltage. It can be mainly found in the circuits in which integrated circuits that require +Vcc and – Vcc are used. Other voltage regulators may require additional components to set the output voltage level, or to assist in the regulation process. 79xx series ICs have built-in protection against a circuit drawing too much current. They have protection against

overheating and short-circuits, making them quite robust in most applications. In some cases, the current-limiting features of the 79xx devices can provide protection not only for the 79xx itself, but also for other parts of the circuit

79XX	Output Voltage	Input Voltage
7905	-5	+7.5
7912	-12	+14
7915	-15	+17
7918	-18	+21

The LM317 is an adjustable 3−terminal positive voltage regulator capable of supplying in excess of 1.5 A over an output voltage range of 1.2 V to 37 V. This voltage regulator is exceptionally easy to use and requires only two external resistors to set the output voltage. Further, it employs internal current limiting, thermal shutdown and safe area compensation, making it essentially blow−out proof. The LM317 serves a wide variety of applications including local, on card regulation. This device can also be used to make a programmable output regulator, or by connecting a fixed resistor between the adjustment and output, the LM317 can be used as a precision current regulator.

In this circuit we are interfacing the fixed voltage regulator 7805 with variable voltage regulator LM371 and we use to display output voltage to the LCD. This can be use for many digital application for testing purpose. This circuit is interface with ATmega16, we are also using the analog feature of AVR to make efficient work. We connect the feedback loop to analog of ATmega16 for measure the output voltage .

Command mention "lcd_init()" is important command to initialize the LCD command from header file.

Command mention "init_adc()"() " is important command to initialize the ADC command from header file.

Command mention "x=read_adc(0)" this command reading the ADC value and assigning to 'x' whose data type is integer.

You can make the voltage display more efficient by reducing the window in if condition and increase more condition for EX we currently mention :-

" if(x>500 && x<600) " instead of this you can make precise by " if(x>500 && x<520)" reducing window

But it will increase your condition and second you have to check the voltage in multimeter to the equivalent ADC reading and assign that Voltage value to that ADC reading that you have measure.

Programming is done in "AVRstudio4 " and header are mention in program are mention at the end. The component are used are given below.

WARNING

Do not give supply more than 5 volt to the ADC port of ATmega16

The header we used having predefine PORT to LCD pin connection on PORT – B as mention below, Inter changing of connection can create malfunctioning of LCD or may not run

Component details

1) AVR ATmega16 IC3
2) 7805 IC2
3) LM371
4) 16x2 LCD
5) LED
6) Crystal 16 MHz
7) 12 DC adaptor
8) 10K Preset
9) Capacitor – 0.1 uf
10) Capacitor – 10 uf
11) 10K pot
12) Resistor – 240 ohm
13) ISP burner

Program

```c
#include<avr/io.h>
#include<lcd/Swits.h>
#include<util/delay.h>
main()
{
        lcd_init();
init_adc();
int x;
while(1)
{
        x=read_adc(0);          // this command reading the ADC value and assigning to 'x'
        if(x>0 && x<100)        // If condition
        {
                lcd_clrscr();
                lcd_goto(0,1);
                lcd_prints("0.5 volt");
                _delay_ms(150);
        }
        if(x>100 && x<200)      // If condition
            {
            lcd_clrscr();
            lcd_goto(0,1);
            lcd_prints("1 volt");
            _delay_ms(150);
            }
```

```c
if(x>200 && x<300)
    {
    lcd_clrscr();
    lcd_goto(0,1);
    lcd_prints("1.5 volt");
    _delay_ms(150);
    }
if(x>300 && x<400)
    {
    lcd_clrscr();
    lcd_goto(0,1);
    lcd_prints("2 volt");
    _delay_ms(150);
    }
if(x>400 && x<500)
    {
    lcd_clrscr();
    lcd_goto(0,1);
    lcd_prints("2.5 volt");
    _delay_ms(150);
    }
if(x>500 && x<600)
    {
    lcd_clrscr();
    lcd_goto(0,1);
    lcd_prints("3 volt");
```

```c
        _delay_ms(150);
    }
if(x>600 && x<700)        // If condition
    {
    lcd_clrscr();
    lcd_goto(0,1);
    lcd_prints("3.5 volt");
    _delay_ms(150);
    }
if(x>700 && x<800)        // If condition
    {
    lcd_clrscr();
    lcd_goto(0,1);
    lcd_prints("4 volt");
    _delay_ms(150);
    }
if(x>800 && x<900)        // If condition
    {
    lcd_clrscr();
    lcd_goto(0,1);
    lcd_prints("4.5 volt");
    _delay_ms(150);
    }
if(x>900 && x<1023)       // If condition
    {
    lcd_clrscr();
```

```
            lcd_goto(0,1);

            lcd_prints("5 volt");

            _delay_ms(150);

            }
      }
}
```

Schismatic Diagram

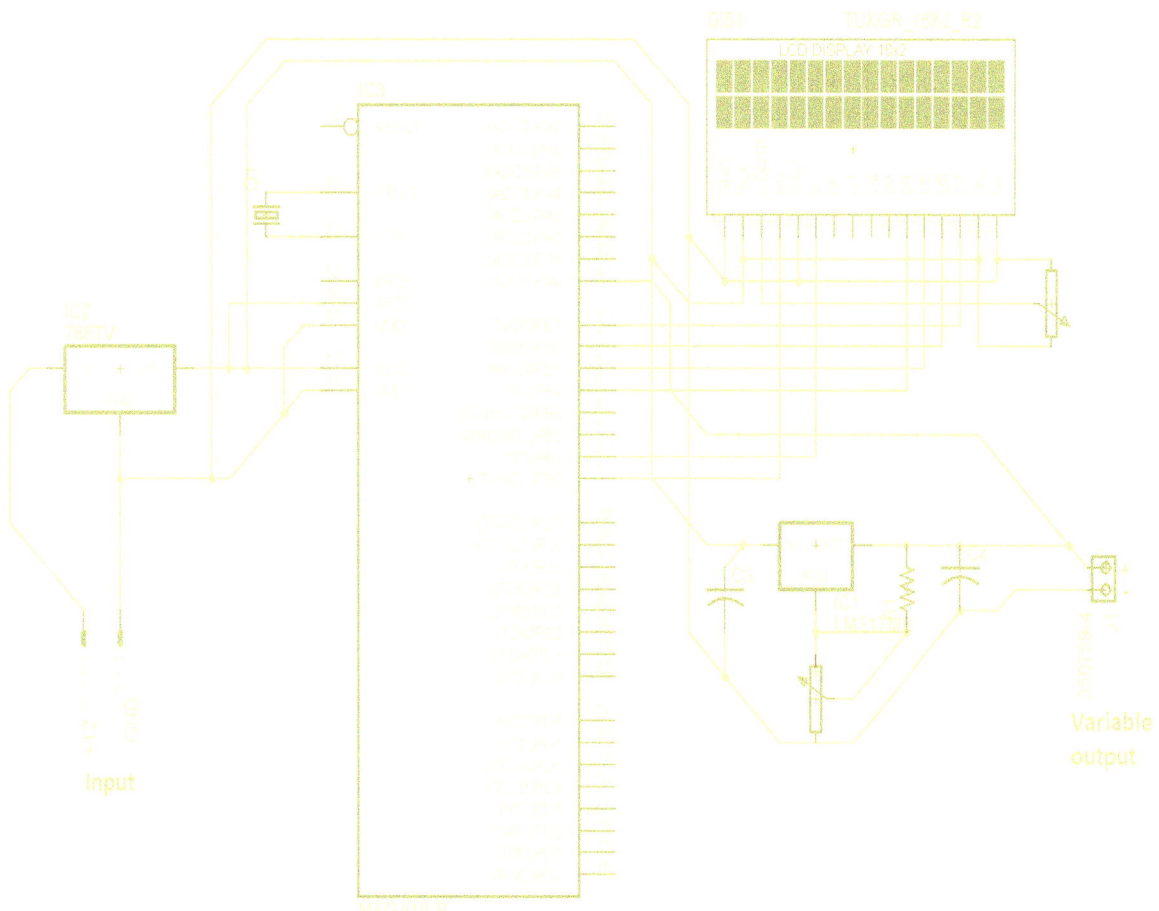

Experiment – 8

Relay Interfacing

In some electronic applications we need to switch or control high voltages or high currents. In these cases we may use electromagnetic or solid state relays. For example, it can be used to control home appliances using low power electronic circuits. Relays are devices which allow low power circuits to switch a relatively high Current/Voltage ON/OFF. For a relay to operate a suitable *pull-in & holding current* should be passed through its coil. Generally relay coils are designed to operate from a particular voltage often its 5V or 12V.

A relay is an electrically operated switch. Many relays use an electromagnet to mechanically operate a switch, but other operating principles are also used, such as solid-state relays. Relays are used where it is necessary to control a circuit by a low-power signal (with complete electrical isolation between control and controlled circuits), or where several circuits must be controlled by one signal. The first relays were used in long distance telegraph circuits as amplifiers: they repeated the signal coming in from one circuit and re-transmitted it on another circuit. Relays were used extensively in telephone exchanges and early computers to perform logical operations.

A type of relay that can handle the high power required to directly control an electric motor or other loads is called a contactor. Solid-state relays control power circuits with no moving parts, instead using a semiconductor device to perform switching. Relays with calibrated operating characteristics and sometimes multiple operating coils are used to protect electrical circuits from overload or faults; in modern electric power systems these functions are performed by digital instruments still called "protective relays".

A simple electromagnetic relay consists of a coil of wire wrapped around a soft iron core, an iron yoke which provides a low reluctance path for magnetic flux, a movable iron armature, and one or more sets of contacts (there are two in the relay pictured). The armature is hinged to the yoke and mechanically linked to one or more sets of moving contacts. It is held in place by a spring so that when the relay is de-energized there is an air gap in the magnetic circuit. In this condition, one of the two sets of contacts in the relay pictured is closed, and the other set is open. Other relays may have more or fewer sets of contacts depending on their function. The relay also has a wire connecting the armature to the yoke. This ensures continuity of the circuit between the moving contacts on the armature, and the circuit track on the printed circuit board (PCB) via the yoke, which is soldered to the PCB.

When an electric current is passed through the coil it generates a magnetic field that activates the armature, and the consequent movement of the movable contact(s) either makes or breaks (depending upon construction) a connection with a fixed contact. If the set of contacts was closed when the relay was de-energized, then the movement opens the contacts and breaks the connection, and vice versa if the contacts were open. When the current to the coil is switched off, the armature is returned by a force, approximately half as strong as the magnetic force, to its relaxed position. Usually this force is provided by a spring, but gravity is also used commonly in industrial motor starters. Most relays are manufactured to operate quickly. In a low-voltage application this reduces noise; in a high voltage or current application it reduces arcing.

When the coil is energized with direct current, a diode is often placed across the coil to dissipate the energy from the collapsing magnetic field at deactivation, which would otherwise generate a voltage spike dangerous to semiconductor circuit components. Such diodes were not widely used before the application of transistors as relay drivers, but soon became ubiquitous as early germanium transistors were easily destroyed by this surge. Some automotive relays include a diode inside the relay case.

If the relay is driving a large, or especially a reactive load, there may be a similar problem of surge currents around the relay output contacts. In this case a snubber circuit (a capacitor and resistor in series) across the contacts may absorb the surge. Suitably rated capacitors and the associated resistor are sold as a single packaged component for this commonplace use.

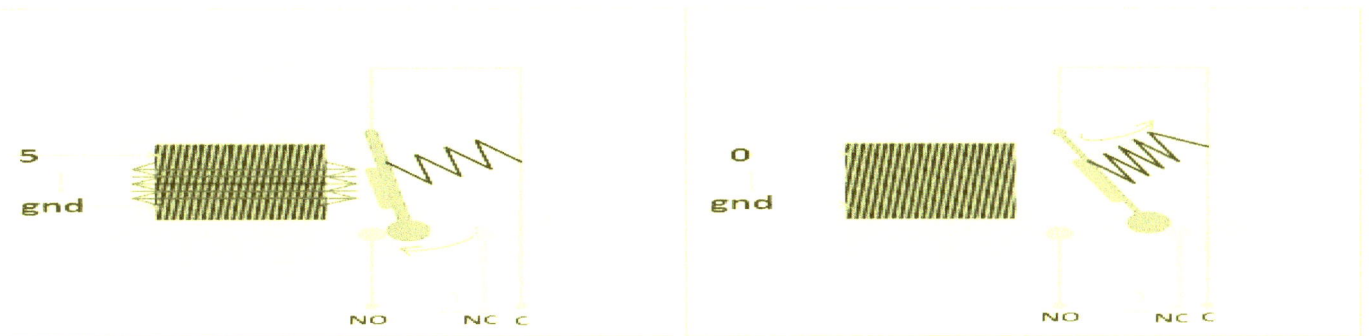

If the coil is designed to be energized with alternating current (AC), some method is used to split the flux into two out-of-phase components which add together, increasing the minimum pull on the armature during the AC cycle.

The relay come with various terminology they are :-

SPST – Single Pole Single Throw. These have two terminals which can be connected or disconnected. Including two for the coil, such a relay has four terminals in total. It is ambiguous whether the pole is normally open or normally closed. The terminology "SPNO" and "SPNC" is sometimes used to resolve the ambiguity.

SPDT – Single Pole Double Throw. A common terminal connects to either of two others. Including two for the coil, such a relay has five terminals in total.

DPST – Double Pole Single Throw. These have two pairs of terminals. Equivalent to two SPST switches or relays actuated by a single coil. Including two for the coil, such a relay has six terminals in total. The poles may be Form A or Form B (or one of each).

DPDT – Double Pole Double Throw. These have two rows of change-over terminals. Equivalent to two SPDT switches or relays actuated by a single coil. Such a relay has eight terminals, including the coil.

Relays are switches, the terminology applied to switches the terminal named as NO, NC and common.

Normally open (NO) contacts connect the circuit when the relay is activated; the circuit is disconnected when the relay is inactive. Normally closed (NC) contacts disconnect the circuit when the relay is activated; the circuit is connected when the relay is inactive. Common is the contact which can shift from NO to NC when the coil is activate or deactivate.

The relay can be interfaced in two ways first by using transistor and second is by using L293D. In this practice we are going to use L293D to interface the relay with Atmega16.

L293D are quadruple high-current half-H drivers. The L293 is designed to provide bidirectional drive currents of up to 1 A at voltages from 4.5 V to 36 V. The L293D is designed to provide bidirectional drive currents of up to 600-mA at voltages from 4.5 V to 36 V. Both devices are designed to drive inductive loads such as relays, solenoids, dc and bipolar stepping motors, as well as other high-current/high-voltage loads in positive-supply applications. All inputs are TTL compatible. Each output is a complete totem-pole drive circuit, with a Darlington transistor sink and a pseudo Darlington source. Drivers are enabled in pairs, with drivers 1 and 2 enabled by 1,2EN and drivers 3 and 4 enabled by 3,4EN. When an enable input is high the associated drivers are enabled, and their outputs are active and in phase with their inputs. When the enable input is low, those drivers are disabled, and their outputs are off and in the high-impedance state. With the proper data inputs, each pair of drivers forms a full-H (or bridge) reversible drive suitable for relay or motor applications.

We are using 2 port of atmega16 PORT-A for giving switch input to make relay on and off and PORT-B to connect relay via L293D. In this practical when switch 1 is trigger logical 5volt is get to the controller which make the PORT-D pin high and coil of relay energized and change the state from normal connected to normal open but when switch 2 is trigger logical 5volt is get to the controller at port pin PA1 which make the PORT-D pin Low and coil of relay de-energized and change the state from normal open to normal connected. The AC appliance we connected to relay we also change the state from ON and OFF as relay change its states.

The line mention " PORTA= 0b11111111; " & " PORTD= 0b00000000; state that we are assigning logic 5 volte and logic 0 volt to port A & port D respectively and every pin in port A & D from PORTA0, D0 – PORTA7, D7 is high and low respectively. We can assign any pin of PORTA to turn on and off the AC appliances.

The line mention "DDRA = 0b11111111" & "DDRD = 0b00000000" This stands for Directive register to be select; the port we are using is output port or input port, where 1 stands for output and 0 stands for input.

Programming is done in "AVRstudio4 " and header are mention in program are mention at the end. The component are used are given below.

WARNING

Do not connect AC supply in digital circuits rather than relay contact points

Component details

1) AVR ATmega16 IC3

2) 7805 IC2

3) Push to ON switch

4) L293D IC1

5) Relay 12v

6) Crystal 16 MHz

7) Adaptor 12v

8) ISP Burner

Program

#include<avr/io.h>

#include<util/delay.h>

main()

{

DDRA = 0b00000000;

DDRD = 0b11111111;

PORTA= 0b1111111;

PORTD = 0b0000000;

    while(1)

    {

        if(PINA == 0b00000001)

            {

                PORTD = 0b00000011;

                _delay_ms(150);

            }

        if(PINA == 0b00000010)

            {

                PORTD = 0b00000000;

                _delay_ms(150);   // delay
            }
        }
    }

Schismatic Diagram

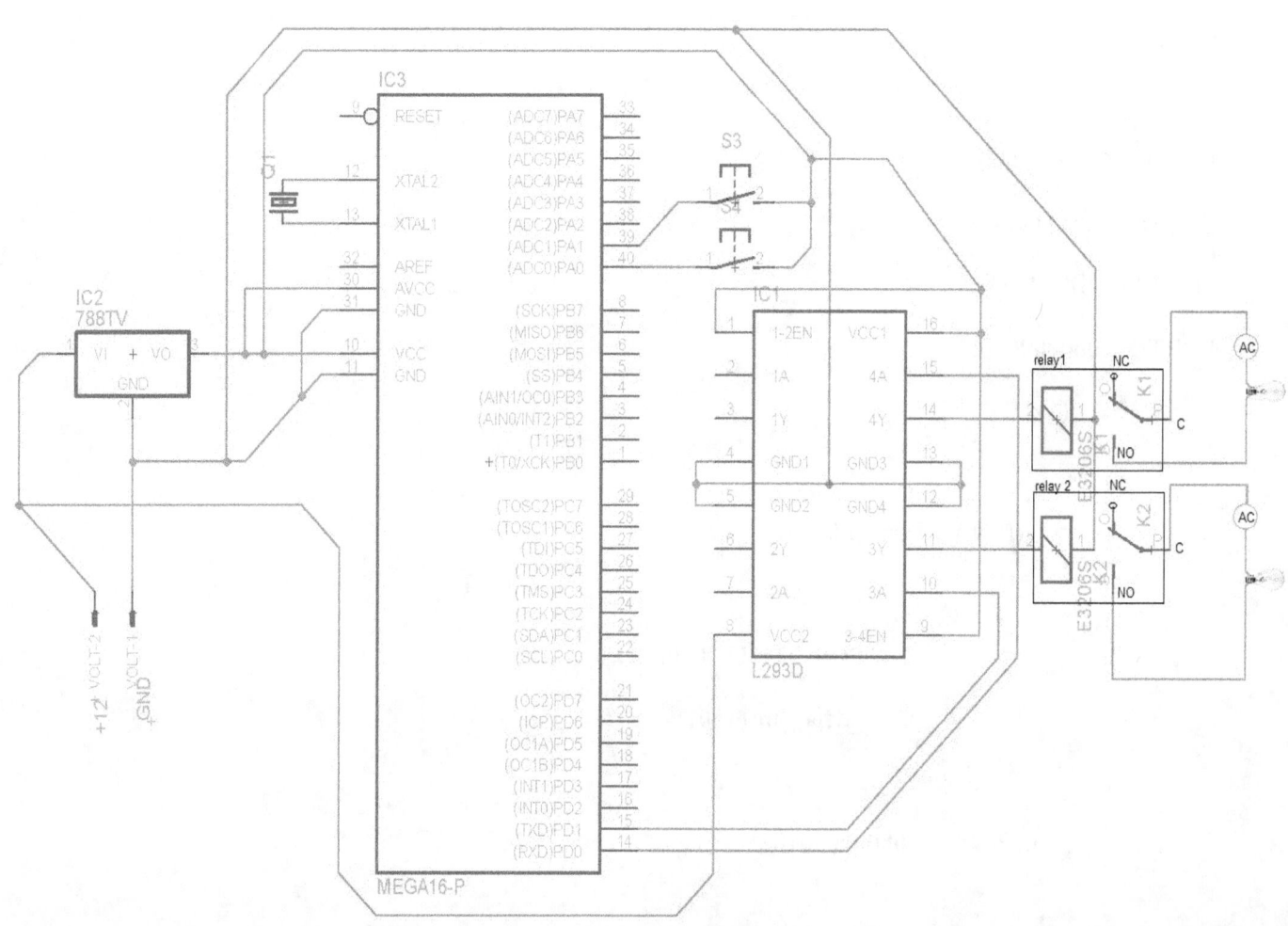

Experiment – 9

Dc motor interfacing

A DC motor is any of a class of electrical machines that converts direct current electrical power into mechanical power. The most common types rely on the forces produced by magnetic fields. Nearly all types of DC motors have some internal mechanism, either electromechanical or electronic, to periodically change the direction of current flow in part of the motor. Most types produce rotary motion a linear motor directly produces force and motion in a straight line.

DC motors were the first type widely used, since they could be powered from existing direct-current lighting power distribution systems. A DC motor's speed can be controlled over a wide range, using either a variable supply voltage or by changing the strength of current in its field windings. Small DC motors are used in tools, toys, and appliances. The universal motor can operate on direct current but is a lightweight motor used for portable power tools and appliances. Larger DC motors are used in propulsion of electric vehicles, elevator and hoists, or in drives for steel rolling mills. The advent of power electronics has made replacement of DC motors with AC motors possible in many applications.

A coil of wire with a current running through it generates an electromagnetic field aligned with the center of the coil. The direction and magnitude of the magnetic field produced by the coil can be changed with the direction and magnitude of the current flowing through it. A simple DC motor has a stationary set of magnets in the stator and an armature with one more windings of insulated wire wrapped around a soft iron core that concentrates the magnetic field. The windings usually have multiple turns around the core, and in large motors there can be several parallel current paths. The ends of the wire winding are connected to a commutator. The commutator allows each armature coil to be energized in turn and connects the rotating coils with the external power supply through brushes. (Brushless DC motors have electronics that switch the DC current to each coil on and off and have no brushes.)

The total amount of current sent to the coil, the coil's size and what it's wrapped around dictate the strength of the electromagnetic field created. The sequence of turning a particular coil on or off dictates what direction the effective electromagnetic fields are pointed. By turning on and off coils in sequence a rotating magnetic field can be created. These rotating magnetic fields interact with the magnetic fields of the magnets (permanent or electromagnets) in the stationary part of the motor (stator) to create a force on the armature which causes it to rotate. In some DC motor designs the stator fields use electromagnets to create their magnetic fields which allow greater control over the motor. Typical brushless DC motors use one or more permanent magnets in the rotor and electromagnets on the motor housing for the stator.

There are various type of motor are present in various voltage from 5 volt to 48 volt they also have different current ratting. The maximum current that can be sourced or sunk from ATmega microcontroller is 15 mA at 5v. But a DC Motor need currents very much more than that and it need voltages 6v, 12v, 24v etc, depending upon the type of motor used and Another problem is that the back emf produced by the motor may affect the proper functioning of the microcontroller. Due to these reasons we can't connect a DC Motor directly to a microcontroller. To overcome this problem the L293D driver IC is used. It is a Quadruple Half H-Bridge driver and it solves the problem completely. You needn't connect any transistors, resistors or diodes. We can easily control the switching of L293D using a microcontroller. There are two IC's in this category L293D and L293. L239D can provide a maximum current of 600mA from 4.5V to 36V while L293 can provide up to 1A under the same input conditions. All inputs of these ICs are TTL compatible and clamp diodes is provided with all outputs. They are used with inductive loads such as relays solenoids, motors etc.

We are using two port for controller the motor to change state of motor reverse and forward and to change ON – OFF state. In this practice we are using PORT-A to give input via switch and PORT-D to connect motor via motor driver L293D.

// condition 1 – Enter in main while loop.

// condition 2 – If button1 condition is satisfied enter in second while loop.

// condition 3 – in the condition there is again while loop, in this loop if button1 is pressed motor starts in clock direction and if button2 is pressed motor starts in anti-clock direction and loop breaks.

// condition 4 – clock direction

// condition 5 – Anti clock direction

// condition 6 – In the main loop button 2 is pressed the motor get stop.

The line mention " PORTA= 0b11111111; " & " PORTD= 0b00000000; state that we are assigning logic 5 volte and logic 0 volt to port A & port D respectively and every pin in port A & D from PORTA0, D0 – PORTA7, D7 is high and low respectively. We can assign any pin of PORTA to turn on and off the Motor and to change the direction of motor.

The line mention "DDRA = 0b11111111" & "DDRB = 0b00000000" This stands for Directive register to be select; the port we are using is output port or input port, where 1 stands for output and 0 stands for input.

Programming is done in "AVRstudio4 " and header are mention in program are mention at the end. The component are used are given below.

Component details

1) AVR ATmega16 IC3

2) 7805 IC2

3) Crystal 16 MHz

4) L293D

5) Button

6) 12V DC motor

7) 12 DC adaptor

8) ISP burner

Program

#include<avr/io.h>

#include<util/delay.h>

main()

{

DDRA = 0b00000000;

DDRD = 0b11111111;

PORTA= 0b1111111;

PORTD = 0b0000000;

while(1)

{

if(PINA == 0b00000001)

{

While(1)

{

_delay_ms(250);

if(PINA == 0b00000010)

{

PORTD = 0b00000010;

_delay_ms(150);

break;

}

if(PINA == 0b00000001)

{

```c
                        PORTD = 0b00000001;         // port set to logical 5 volt to ON LED

                        _delay_ms(150);             // delay


                    break;
                    }

                }
            }
        if(PINA == 0b00000010)                      // condition 6
            {
                PORTD = 0b00000000;                 // port set to logical 5 volt to ON LED
                _delay_ms(150);                     // delay
            }
        }


}
```

Schismatic Diagram

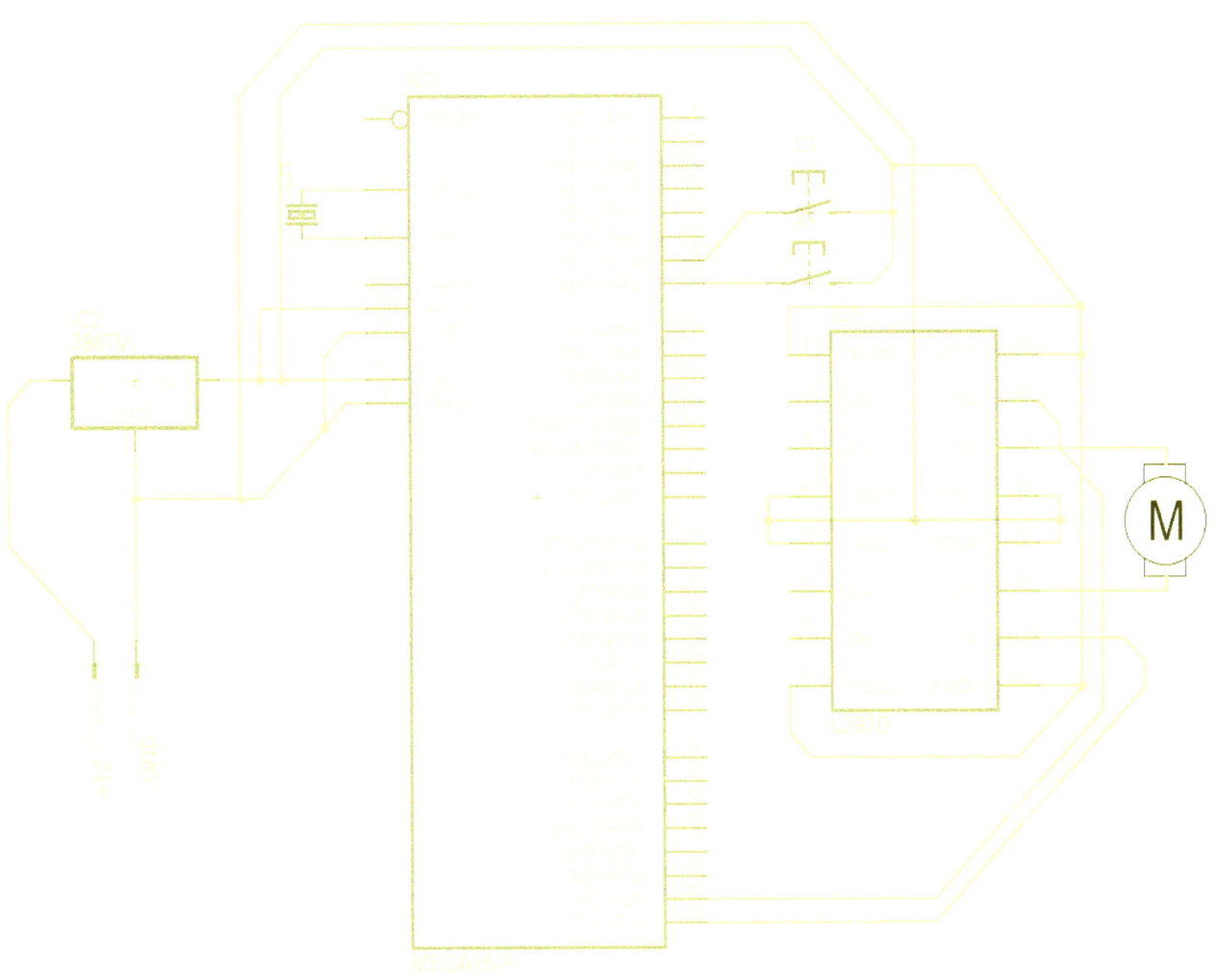

Experiment – 10

Propeller Display

This practice is started with a simple principle which is frequently encountered in our everyday life, which is Persistence of Vision. This phenomenon makes one feel fast moving/changing objects to appear continuous. A television is a common example in which image is re-scanned every 25 times, thereby appear continuous.

Further, a glowing objects if rotated in a circle at fast speed, it shows a continuous circle. By modifying this basic idea, 8 LEDs can be rotated in a circle, showing 8 concentric circles. But if these LEDs are switched at precise intervals, a steady display pattern can be shown.

The timing parameters are very important in this project. A small change in time delay can cause problem. Means it will not display properly which we want to display. so the delay should be proper between characters.

As shown in to the figure the time is not properly set. so it is not displaying properly. At least 10 to 15 tests are required to set the proper delay and to display characters properly. But once time is set properly, then the problem will solve. The string can be changed according to requirement. Display can be easily changed just by changing their codes.

It's simple circuit of LED blinker just difference is the circuit assembly is placed on rotating propeller which is attached to motor.

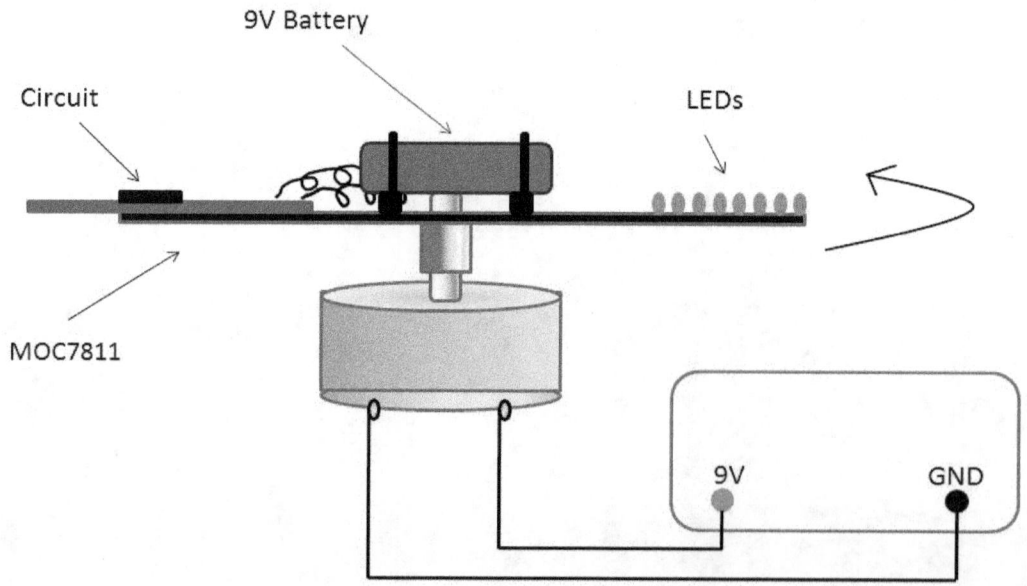

The assembly can look like as shown in Fig.

There are various parameter such as battery fitted at center otherwise it can create vibration and circuit may be dislocate in rotary moment. LED should place at equal distance from each other and mounted tightly that they won't create problem of dislocation in rotary moment and the circuit should also mounted properly.

In this circuit we are using PORT-A to connect LED to display word "Hello"and the circuit complete based on delay first we have to check the delay according to the propeller we are designing of such diameter, diameter increase the rotation of propeller also increases.

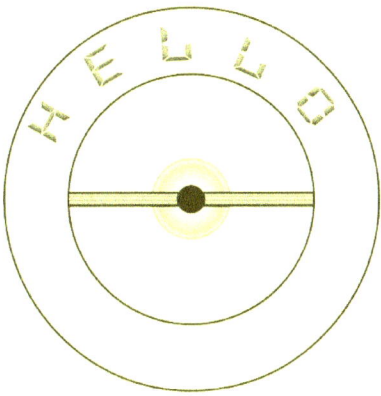

The best idea to decide the word we have to print on propeller just follow few step -> first open "Word Doc" and after insert table of row 8 and X number column and mark the number in box to make pattern.

PA0	0		0		0	0	0	0			0			0	0	0
PA1	0		0		0			0			0			0		0
PA2	0		0		0			0			0			0		0
PA3	0	0	0	0	0	0		0			0			0		0
PA4	0		0		0			0			0			0		0
PA5	0		0		0			0			0			0		0
PA6	0		0		0	0	0	0	0	0	0	0	0	0	0	0

Component details

1) AVR ATmega16 IC3

2) 7805 IC2

3) Crystal 16 MHz
4) LED
5) 9v battery
6) 12V DC motor
7) 12 DC adaptor
8) ISP burner

Program

```c
#include<avr/io.h>
#include<util/delay.h>
main()
{
DDRA = 0b11111111;
PORTA= 0b00000000;
    while(1)
    {
        PORTA= 0b01111111;   // port set to logical 5 volt " Start – H "
        _delay_ms(3);        // delay
        PORTA= 0b00000000;   // port set to logical 0 volt "null"
        _delay_ms(3);        // delay
        PORTA= 0b00001000;   // port set to logical 5 volt " line – H "
        _delay_ms(6);        // delay
        PORTA= 0b00000000;   // port set to logical 0 volt "null"
        _delay_ms(3);        // delay
        PORTA= 0b01111111;   // port set to logical 5 volt " line – H "
```

```
_delay_ms(3);
PORTA= 0b00000000;                " Finish - H"
_delay_ms(9);
PORTA= 0b01111111;                " Start – E "
_delay_ms(3);
PORTA= 0b00000000;                "null"
_delay_ms(3);
PORTA= 0b01001001;                " line – E "

_delay_ms(5);
PORTA= 0b00000000;                "null"
_delay_ms(3);
PORTA= 0b01000001;                " Finish - E "
_delay_ms(3);
PORTA= 0b00000000;                "null"
_delay_ms(9);
PORTA= 0b01111111;                " Start - L "
_delay_ms(3);
PORTA= 0b00000000;                "null"
_delay_ms(3);
PORTA= 0b01000000;                " line – L "
_delay_ms(5);
PORTA= 0b00000000;                "finish - L"
_delay_ms(9);
PORTA= 0b01111111;                " start - L "
_delay_ms(3);
PORTA= 0b00000000;                "null"
```

```c
        _delay_ms(3);            // delay
        PORTA= 0b01000000;       // port set to logical 5 volt " line – L "
        _delay_ms(5);            // delay
        PORTA= 0b00000000;       // port set to logical 0 volt "finish - L"
        _delay_ms(9);            // delay
        PORTA= 0b00111110;       // port set to logical 5 volt " Start - O "
        _delay_ms(3);            // delay
        PORTA= 0b00000000;       // port set to logical 0 volt "null"


        _delay_ms(3);            // delay
        PORTA= 0b01000001;       // port set to logical 5 volt " line – O"
        _delay_ms(5);            // delay
        PORTA= 0b00000000;       // port set to logical 0 volt "null"
        _delay_ms(3);            // delay
        PORTA= 0b00111110;       // port set to logical 5 volt " Finish - O "
        _delay_ms(3);            // delay
        PORTA= 0b00000000;       // port set to logical 0 volt "null"
        _delay_ms(3);            // delay
    }
}
```

Schismatic Diagram

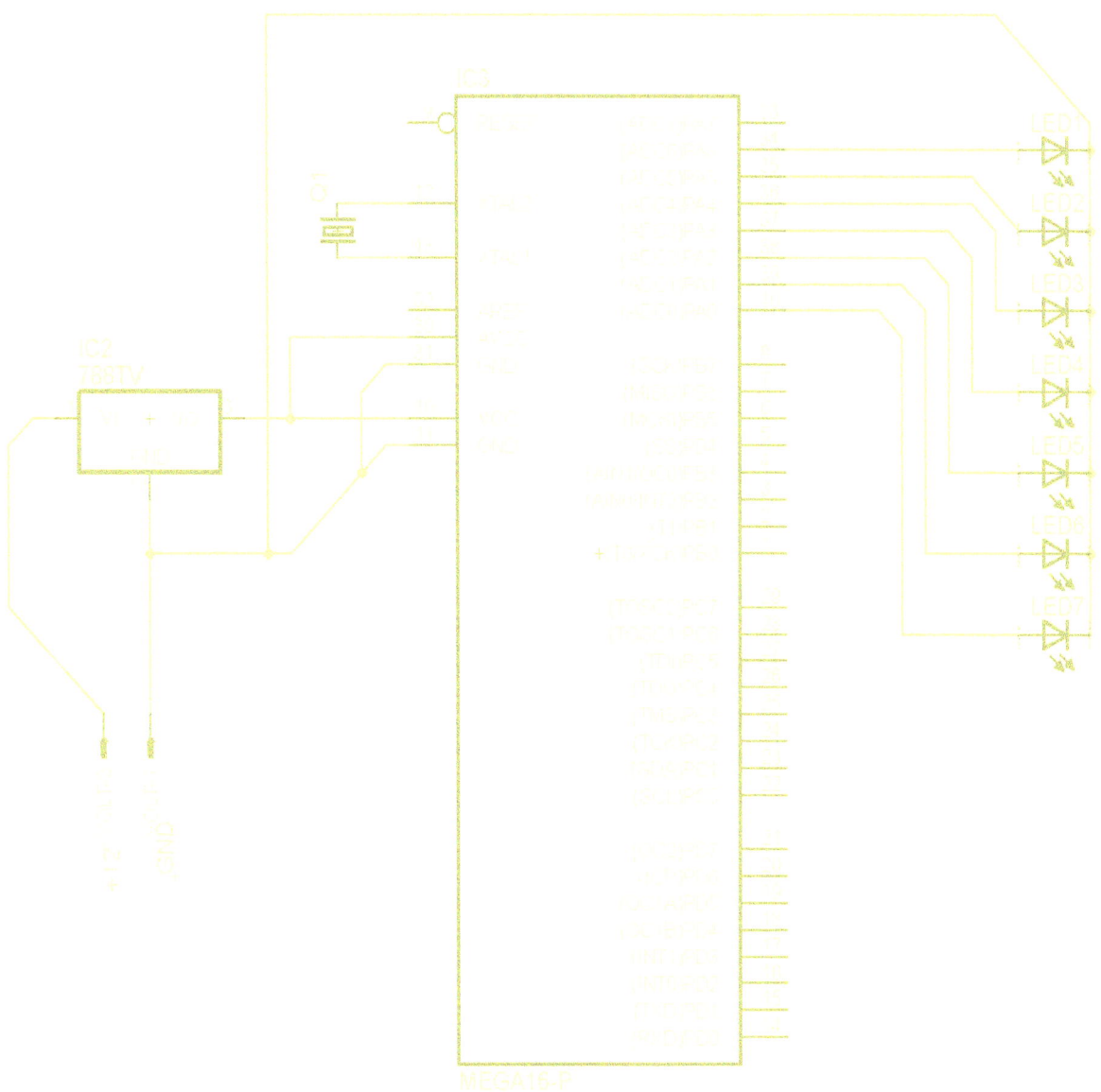

Experiment – 11

Fire detector

In this practical we are going to detect the fire using photo-diode, we are using the presence of infrared in the fire to detect the fire. It uses a photodiode as the fire sensor and blink LED immediately on sensing a spark or fire. The circuit exploits the photovoltaic property of the photodiodes to sense the fire. The photodiode used in infrared detectors generate a photo voltage proportional to the incident light rays or infrared rays falling on it. Typically, 1V is produced in the photodiode when it is forward biased by accepting the photons. Here the passive infrared rays from the spark or fire are used to activate the photodiode to generate the photo voltage.

The photo voltage is very small, hence we are using LM234 opamp to amplify the data signal to 4.5 volts. LM234 consist of four independent high-gain frequency-compensated operational amplifiers that are designed specifically to operate from a single supply over a wide range of voltages. Operation from split supplies also is possible if the difference between the two supplies is 3 V to 32 V (3 V to 26 V for the LM2902 device), and VCC is at least 1.5 V more positive than the input common-mode voltage. The low supply-current drain is independent of the magnitude of the supply voltage.

We are using ADC to detect the signal coming from opamp, this signal is to be sampled and gets the efficient equivalent reading on 16*2 LCD display. This reading is also use to detect the intensity of fire reading can get vary after fire gets detected according to distance.

A photodiode is a semiconductor device that converts light into current. The current is generated when photons are absorbed in the photodiode. A small amount of current is also produced when no light is present. Photodiodes may contain optical filters, built-in lenses, and may have large or small surface areas. Photodiodes usually have a slower response time as their surface area increases.

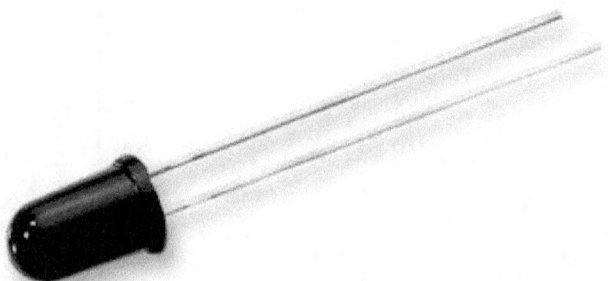

Photo diode   Fig

In this experiment we use ATmega16 due to low flash requirement. In this we use PORT-A as input port which is connected to sensor via opamp, and PORT-B as output port which is connected to LCD. And LED connected to Port-D and We are giving logical 5 volt to PORTD to glow

Working :

The circuit should connect as shown in schematic and program should be as follows.

When we spark fire in front of Photo diode the small amount of current start flow from biasing. This voltage is too small to detect in microcontroller hence we are amplifying this signal using opamp LM324 , this is basically work as comparator to compare the voltage with reference we applied and gives the output. The output comes from opamp is given to analog pin of ATmega16, controller process and give the equivalent output on display.

In this practice we have to read the analog value display on the LCD and put in program. Test the analog value in front fire and note down "x" also note down without fire "y". The reading always in integer from 0-1024.

Line mention this is the condition display on LCD when there is no fire in front of photo diode.

Line mention this is the condition display on LCD when there is fire in front of photo diode.

Programming is done in "AVRstudio4 " and header are mention in program are mention at the end. The component are used are given below.

WARNING

Do not burn photo diode, keep fire distance from photodiode. The fire intensity decides the detection sensitive of fire. If the data signal from fire is less sensitive than only use Opamp in between as shown in fir 11.1

Expose to sunrays may be work malfunction, its only indoor use.

Component details

1) AVR ATmega16 IC3

2) 7805 IC2

3) LM324

4) Photo diode

5) LCD 16x2

6) 10 k preset - 2

7) LED

8) Crystal 16 MHz

9) 9 volt battery cap

10) ISP burner

Program

```c
#include<avr/io.h>
#include<swits.h>
#include<util/delay.h>
main()
{
    lcd_init();
    init_adc();
    int x;
    DDRD=0b11111111;
    PORTD=0b00000000;
while(1)
    {
        x=read_adc(1);
            lcd_clrscr();
                lcd_goto(1,1);
                lcd_printi(x);
                _delay_ms(100);
```

```c
        if(x >= 1024)
        {
                lcd_clrscr();
                lcd_goto(1,1);
                lcd_prints("Fire not detected");
                _delay_ms(100);
        }



if(x <= 950)
        {
                lcd_clrscr();
                lcd_goto(1,1);
                lcd_prints("Fire detected");
                _delay_ms(100);
                PORTD =0b11111111;
                _delay_ms(500);
                PORTD=0b00000000;
        }

    }


}
```

Schematic Diagram

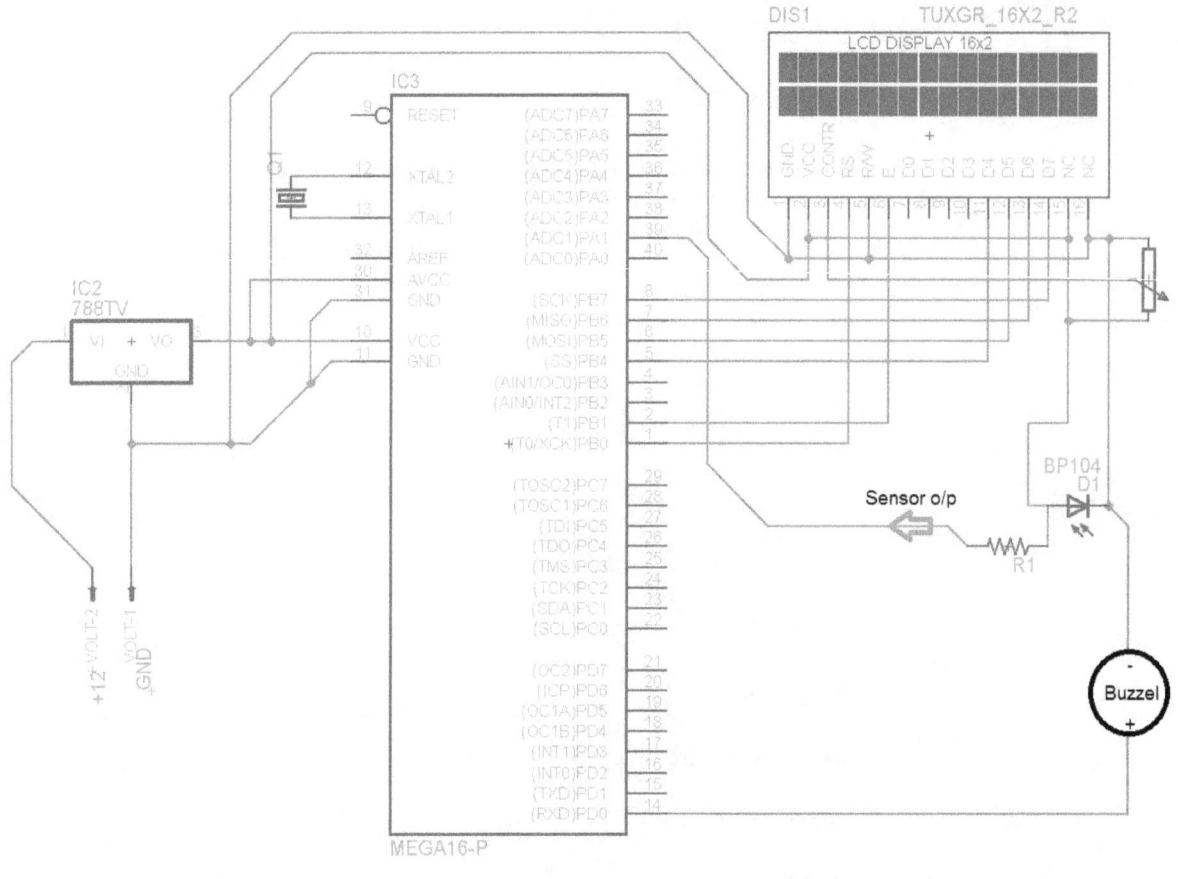

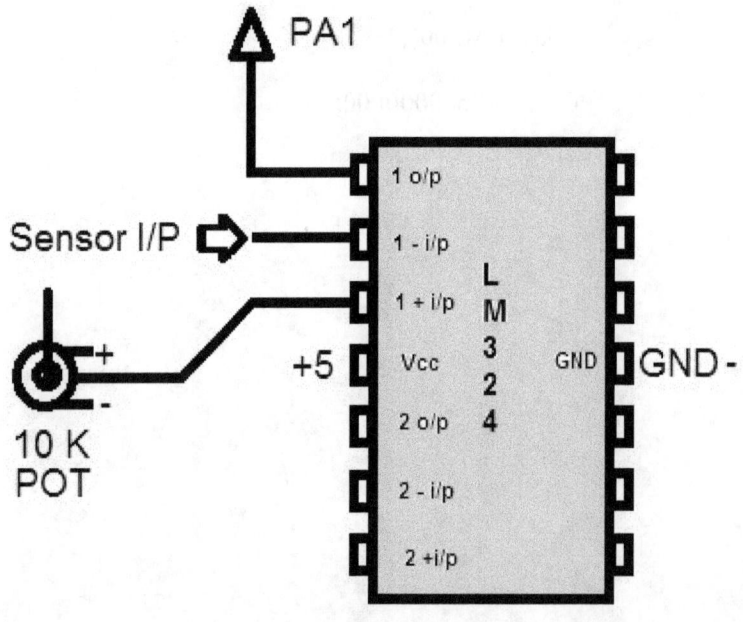

Fig 11.1

Experiment – 12

Day light detector

In this practice we are going to detect day and night. This practical is also almost similar to previous practical. We are using Photodiode to detect the presence of day. A photodiode, used as a photo detector, generates current in the circuit when light incidents on it. This circuit uses the photodiode in reverse bias mode with resistor . This resistor does not allow too much current to flow through the photodiode in case a large amount of light falls on the detector infrared from sun. Initially when no light falls on the photodiode, it results in high potential at the inverting input of a comparator of LM324. When light falls on the photodiode, it allows current to flow through the diode, and thus drops the voltage across it.

Basically we are collection the infrared from the sun light to detect day or night. Sunlight is a portion of the electromagnetic radiation given off by the Sun, in particular infrared, visible, and ultraviolet light. On Earth, sunlight is filtered through Earth's atmosphere, and is obvious as daylight when the Sun is above the horizon. When the direct solar radiation is not blocked by clouds, it is experienced as sunshine, a combination of bright light and radiant heat. When it is blocked by the clouds or reflects off other objects, it is experienced as diffused light. The World Meteorological Organization uses the term "sunshine duration" to mean the cumulative time during which an area receives direct irradiance from the Sun of at least 120 watts per. But we do not affect by cloudy condition because visible sun rays contain ample of infrared to detect by ADC.

Working :

The circuit should connect as shown in schematic and program should be as follows.

When we keep circuit in sun the small amount of current start flow from biasing. This voltage is given to the ADC of ATmega16 to get ADC reading, basically photo diode detect the infrared present in sun and give equivalent output ADC. The output is given to analog pin of ATmega16, controller it process and give the equivalent output on display.

In this practice we have to read the analog value display on the LCD and put in program. Test the analog value in front fire and note down "x" also note down without fire "y". The reading always in integer from 0-1024.

Line mention   this is the condition display on LCD in Night time.

Line mention   this is the condition display on LCD in Day time.

Programming is done in "AVRstudio4 " and header are mention in program are mention at the end. The component are used are given below.

WARNING

This is basic circuit to detect day using infrared present in sun and they can be vary according to sun intensity so to make efficient output test the reading accordingly.

Component details

1) AVR ATmega16 IC3
2) 7805 IC2
3) LM324
4) Photo diode
5) LCD 16x2
6) 10 k preset - 2
7) LED
8) Crystal 16 MHz
9) 9 volt battery cap
10) ISP burner

Program

#include<avr/io.h>

#include<swits.h>

#include<util/delay.h>

main()

{

    lcd_init();

    init_adc();

    int x;

    while(1)

    {

        x=read_adc(1);

        lcd_clrscr();

        lcd_goto(1,1);

        lcd_printi(x);

        _delay_ms(100);

        if(x >= 1024)   // condition is night

        {

            lcd_clrscr();

            lcd_goto(1,1);

            lcd_prints("Night");

            _delay_ms(100);

        }

```c
            if(x <= 700) // condition in day
                {
                    lcd_clrscr();
                    lcd_goto(1,1);
                    lcd_prints("Day");
                    _delay_ms(100);
                }


        }


    }
```

Schismatic Diagram

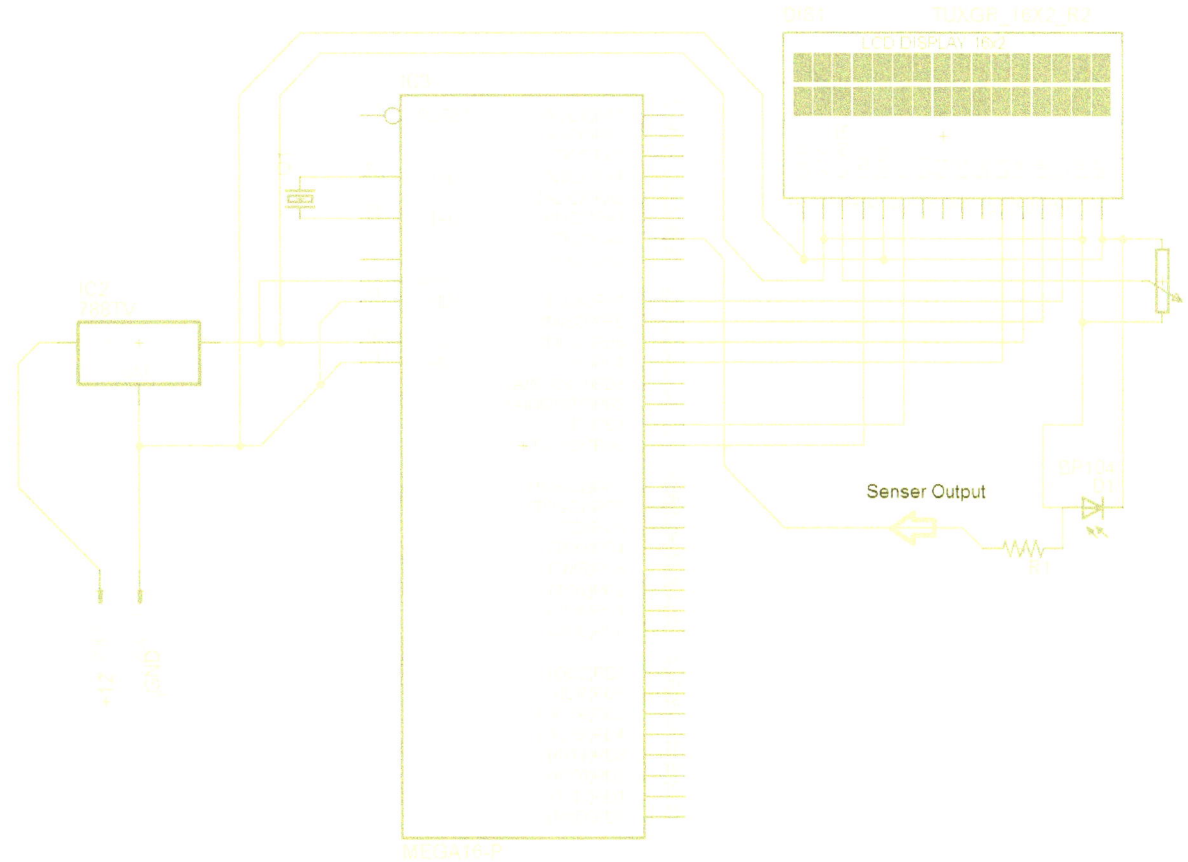

Experiment – 13

Auto day shade light

In this practice we are going to learn how to on automatic light in night using LDR. This is simple practice we had done before just we are going to integrate relay with above circuit.

Light-dependent resistor (LDR) or photocell is a light-controlled variable resistor. The resistance of a photo resistor decreases with increasing incident light intensity; in other words, it exhibits photoconductivity. A photo resistor can be applied in light-sensitive detector circuits, and light- and dark-activated switching circuits.

A photo resistor is made of a high resistance semiconductor. In the dark, a photo resistor can have a resistance as high as several mega ohms (MΩ), while in the light, a photo resistor can have a resistance as low as a few hundred ohms. If incident light on a photo resistor exceeds a certain frequency, photons absorbed by the semiconductor give bound electrons enough energy to jump into the conduction band. The resulting free electrons (and their hole partners) conduct electricity, thereby lowering resistance. The resistance range and sensitivity of a photo resistor can substantially differ among dissimilar devices. Moreover, unique photo resistors may react substantially differently to photons within certain wavelength bands.

A photoelectric device can be either intrinsic or extrinsic. An intrinsic semiconductor has its own charge carriers and is not an efficient semiconductor, for example, silicon. In intrinsic devices the only available electrons are in the valence band, and hence the photon must have enough energy to excite the electron across the entire band gap. Extrinsic devices have impurities, also called dopants, added whose ground state energy is closer to the conduction band; since the electrons do not have as far to jump, lower energy photons (that is, longer wavelengths and lower frequencies) are sufficient to trigger the device. If a sample of silicon has some of its atoms replaced by phosphorus atoms (impurities), there will be extra electrons available for conduction. This is an example of an extrinsic semiconductor.

Working

A light dependent resistor works on the principle of photo conductivity. In day time when light falls on the LDR it has different reading in analog because it passes variable voltage depend on interior resistance produce, but in night time there is absence of light which result in produce high resistance and passes les voltage it is about negligible. This condition is completely satisfy to detect the night and trigger the relay to On state to turn on lights. In this practice we are using AVR atmega16 as we are requiring less flash memory.

Programming is done in "AVRstudio4 " and header are mention in program are mention at the end. The component are used are given below.

WARNING

This is basic circuit to detect day using infrared present in sun and they can be vary according to sun intensity so to make efficient output test the reading accordingly.

Component details

1) AVR ATmega16 IC3
2) 7805 IC2
3) LM324
4) Photo diode
5) LCD 16x2
6) 10 k preset – 2
7) L293D
8) LDR
9) Relay 12V
10) Crystal 16 MHz
11) 9 volt battery cap
12) ISP burner

Program

#include<avr/io.h>

#include<util/delay.h>

#include<swits.h>

main()

{

    lcd_init();   init_adc();

```c
DDRD = 0b11111111; PORTD=0b00000000;

int x;

while(1)
{
    x=read_adc(1);
    if(x >= 900)   // condition in night
    {
        lcd_clrscr();
        lcd_goto(1,1);
        lcd_prints("Night");
        _delay_ms(100);
        PORTD=0b11111111;   //light On (ac appliance)
    }
    Else
    {
        PORTD=0b00000000;      //light off (ac appliance)
    }
}
}
```

Schismatic Diagram

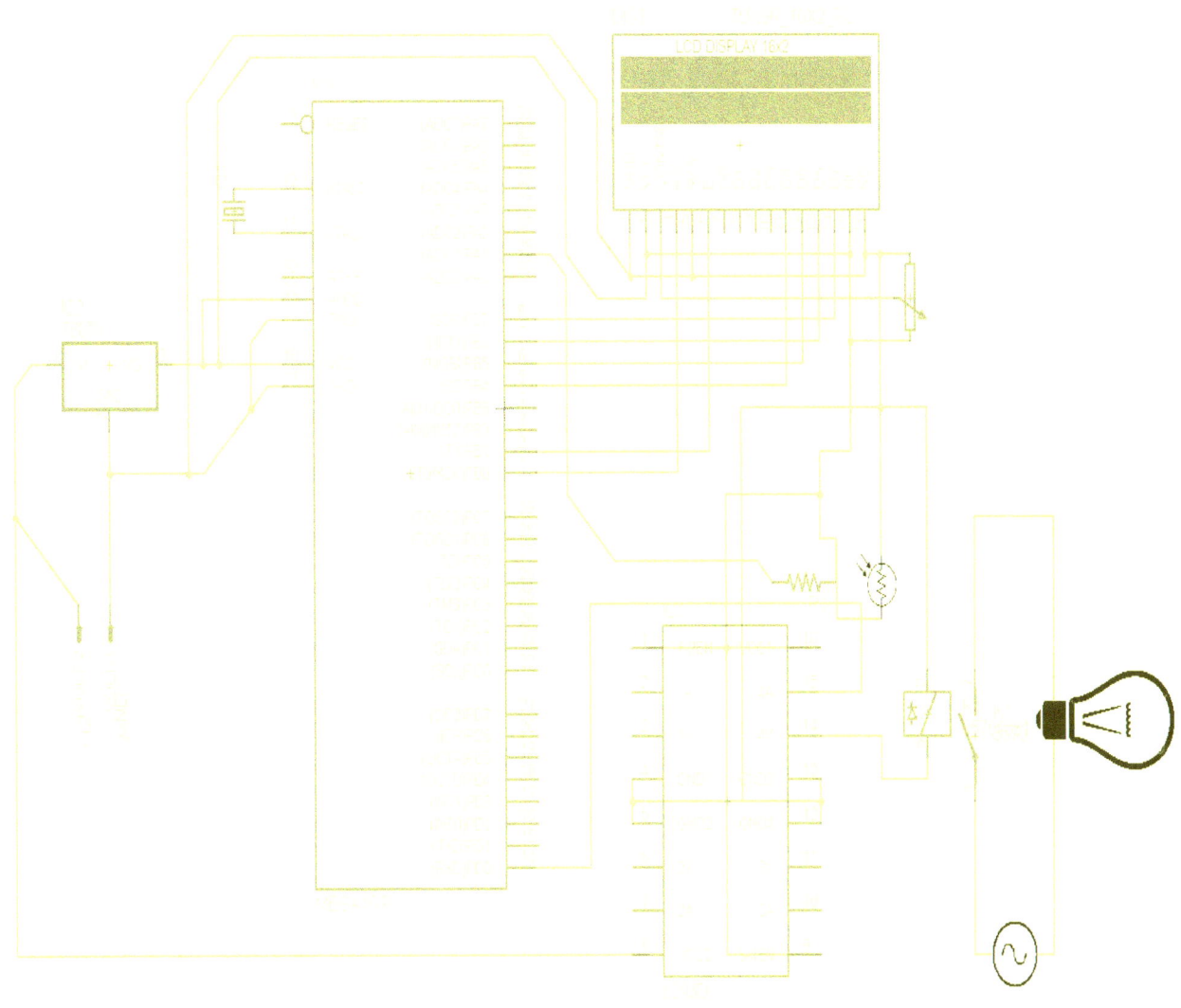

Experiment – 14

Obstacle detector

The basic concept of an Infrared Sensor which is used as Obstacle detector is to transmit an infrared signal, this infrared signal bounces from the surface of an object and the signal is received at the infrared receiver. There are five basic elements used in a typical infrared detection system: an infrared source, a transmission medium, optical component, infrared detectors or receivers and signal processing. Infrared lasers and Infrared LED's of specific wavelength can be used as infrared sources. The three main types of media used for infrared transmission are vacuum, atmosphere and optical fibers. Optical components are used to focus the infrared radiation or to limit the spectral response. Optical lenses made of Quartz, Germanium and Silicon are used to focus the infrared radiation. Infrared receivers can be photodiodes, phototransistors etc. some important specifications of infrared receivers are photosensitivity, selectivity and noise equivalent power. Signal processing is done by amplifiers as the output of infrared detector is very small.

Infrared sensors can be passive or active. Passive infrared sensors are basically Infrared detectors. Passive infrared sensors do not use any infrared source and detects energy emitted by obstacles in the field of view. They are of two types: quantum and thermal. Thermal infrared sensors use infrared energy as the source of heat and are independent of wavelength. Thermocouples, pyroelectric detectors and bolometers are the common types of thermal infrared detectors.

Quantum type infrared detectors offer higher detection performance and are faster than thermal type infrared detectors. The photosensitivity of quantum type detectors is wavelength dependent. Quantum type detectors are further classified into two types: intrinsic and extrinsic types. Intrinsic type quantum detectors are photoconductive cells and photovoltaic cells.

Active infrared sensors consist of two elements: infrared source and infrared detector. Infrared sources include an LED or infrared laser diode. Infrared detectors include photodiodes or phototransistors. The energy emitted by the infrared source is reflected by an object and falls on the infrared detector.

IR Transmitter

Infrared Transmitter is a light emitting diode (LED) which emits infrared radiations. Hence, they are called IR LED's. Even though an IR LED looks like a normal LED, the radiation emitted by it is invisible to the human eye.

There are different types of infrared transmitters depending on their wavelengths, output power and response time. A simple infrared transmitter can be constructed using an infrared LED, a current limiting resistor and a power supply.

IR Receiver

Infrared receivers are also called as infrared sensors as they detect the radiation from an IR transmitter. IR receivers come in the form of photodiodes and phototransistors. Infrared Photodiodes are different from normal photo diodes as they detect only infrared radiation. The picture of a typical IR receiver or a photodiode is shown below.

Different types of IR receivers exist based on the wavelength, voltage, package, etc. When used in an infrared transmitter – receiver combination, the wavelength of the receiver should match with that of the transmitter.

Basic concept to construct Obstruct detector.

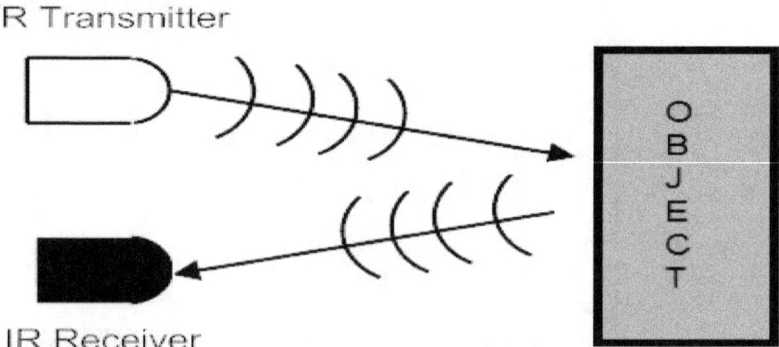

When the IR transmitter emits radiation, it reaches the object and some of the radiation reflects back to the IR receiver. Based on the intensity of the reception by the IR receiver, the output of the sensor is defined. The arrangement should be in such condition that reflection angel should be accepted as shown above. We are using ATmega16 due to require of less flash and interfacing the sensor with controller.

Programming is done in "AVRstudio4 " and header are mention in program are mention at the end. The component are used are given below.

WARNING

Sensor will malfunction in outdoor condition.

Component details

1) AVR ATmega16 IC3

2) 7805 IC2

3) Photo diode

4) Infrared diode

5) 10 k preset – 2

6) L293D

7) Relay 12V

8) Crystal 16 MHz

9) 9 volt battery cap

10) Resistor R1- 10K, R2- 1K

11) ISP burner

Program

#include<avr/io.h>

#include<util/delay.h>

#include<swits.h>

```c
main()
{
    lcd_init();
    init_adc();
    int x;
    while(1)
    {
        x=read_adc(1);
        if(x <= 950)  // condition in night
        {
            lcd_clrscr();
            lcd_goto(1,1);
            lcd_prints("Object Detect");
            _delay_ms(100);
            PORTD=0b11111111;   //buzzel on
        }
        Else
        {
            PORTD=0b00000000;        //buzzel off
        }
    }
}
```

Schismatic Diagram

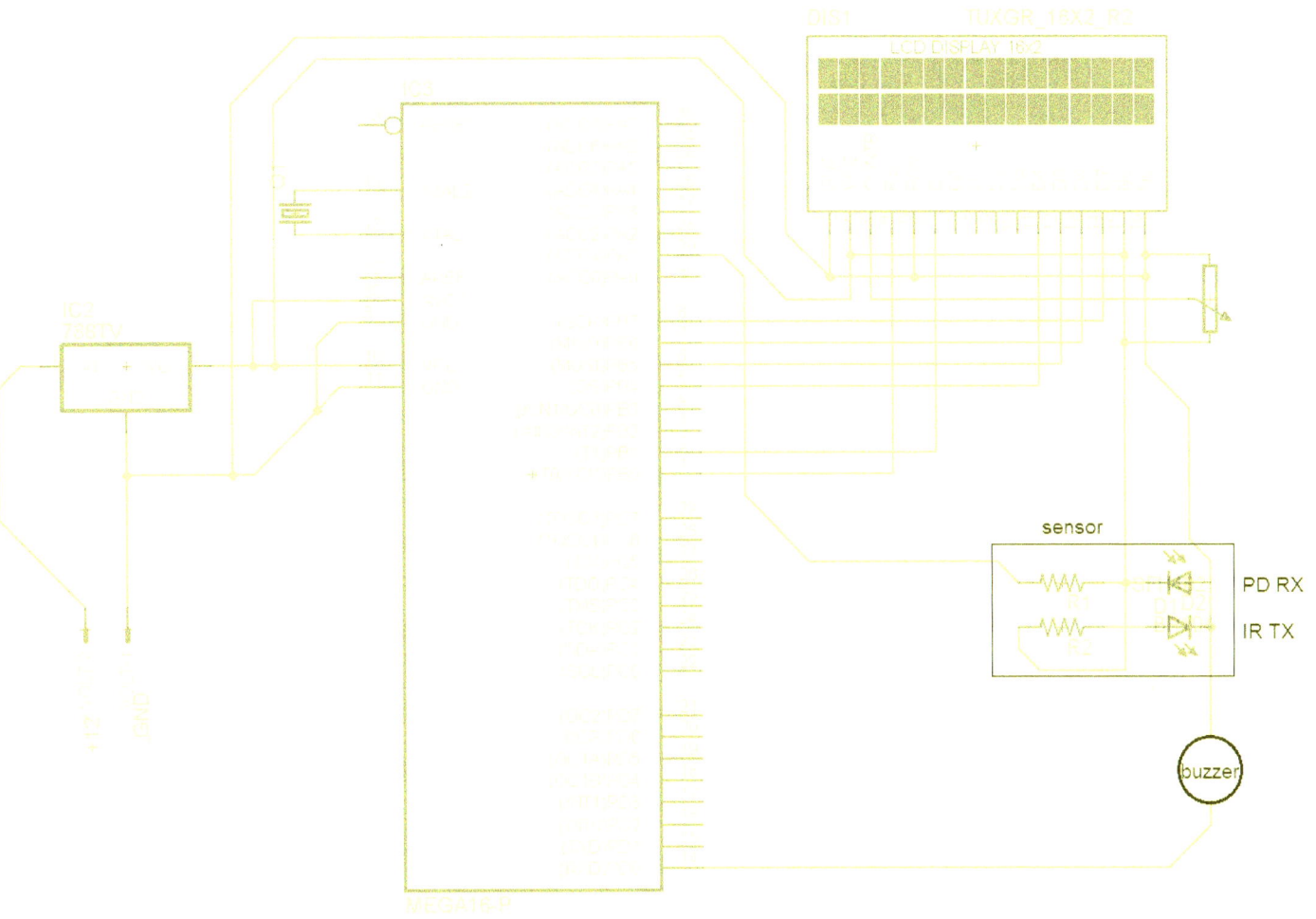

Experiment – 15

Introduction to PWM

PWM stands for Pulse Width Modulation and is the method to produce variable voltages using digital means. Typically, variable voltages come from analog circuits, and digital circuits produce only two voltages, the high (5v, 3.3v, 1.8v, etc.) or low (0v). So how it is possible that digital circuits can produce a voltage that is between the high and the low voltages? If you bring a digital signal up and down, in a consistent manner, you will get a proportion of the voltage between the high and low voltage. Imagine if a digital signal was pulsed high (5v) and low (0v) evenly, say the signal was in the high state for 1 microsecond and in the low state for 1 microsecond, add a capacitor to smooth the signal, the voltage would measure 2.5 volts. Now, change the high voltage in the high state for 9 microseconds and in the low state for 1 microseconds, the voltage would measure 90% of 5 volts, or 5v x .9 = 4.5 volts. The 90% is significant because the duty cycle is represented as a percentage (%). The applications associated with PWM could be: the control of motors, sound output, dimming LEDs, and producing approximated analog waveforms.

In this practical we are performing the PWM testing on LED, because PWM has many requirements and specifications that are very important to insure that you are outputting a PWM signal that will be accepted by the device that is receiving it. The device receiving the PWM that is being outputted by your microcontroller will require the PWM to be at a particular frequency. The period of the PWM is what creates the frequency and it is represented as a length of time. The period is where the digital signal is held high and then goes low and the proportion within this period is the duty cycle. The period is selected initially and doesn't change. The longer the period, the slower the frequency and the sorter the period, the faster the frequency. The frequency of the PWM is how many of these periods can fit within one second. If the period is 1 millisecond long, then the frequency would be 1 kHz, or 1000 Hz, or 1000 times per second.

There are a couple types of PWM, including phase correct where the pulse happens right in the middle of the period, and standard PWM where the pulse happens at the end of the period. The duty cycle, as stated above, is the percentage that the pulse is high within the period. For instance, a 50% duty cycle would be half of the high voltage level.

If we have a quick enough PWM, you can also make analog waveforms of almost any kind. By varying the duty cycle at each period, you could essentially draw the waveform and have it output that way, but the waveform will appear a bit stair like, rather than a perfect waveform. The smaller the periods, or higher the frequency, the better and more smooth the waveforms can be.

In this practice we are using ATmega16 because it has inbuilt function of PWM on predefined pin on PORT D&B.

We are interfacing LED on PORT B and switching ON – OFF smoothly using PWM.

Programming is done in "AVRstudio4" and header are mention in program are mention at the end. The component are used are given below.

WARNING

The header file used "PWM.h" is important header where port pin is predefine for LED. Header is mention at the end of practical in 'required appendix'

Component details

1) AVR ATmega16 IC3
2) 7805 IC2
3) Crystal 16 MHz
4) LED
5) 9 volt battery cap
6) ISP burner

Program

#include<avr/io.h>

```c
#include<util/delay.h>
#include<swits.h>
#include<nir/PWM.h>

void main()
{
uint8_t brightness=0;      // Unsigned integers of 8 bits
InitPWM();                 // initialize PWM from header
while(1)
{
        for(brightness=0;brightness<255;brightness++) // for loop for increasing LED light
        {
                SetPWMOutput(brightness);
                        _delay_ms(250);            // this delay justify the smoothness of glowing up LED
        }
        for(brightness=255;brightness>0;brightness--) // for loop for decreasing LED light
        {
                SetPWMOutput(brightness);
                _delay_ms(250);                    // this delay justify the smoothness of glowing down LED
        }
}       }
```

Schismatic Diagram

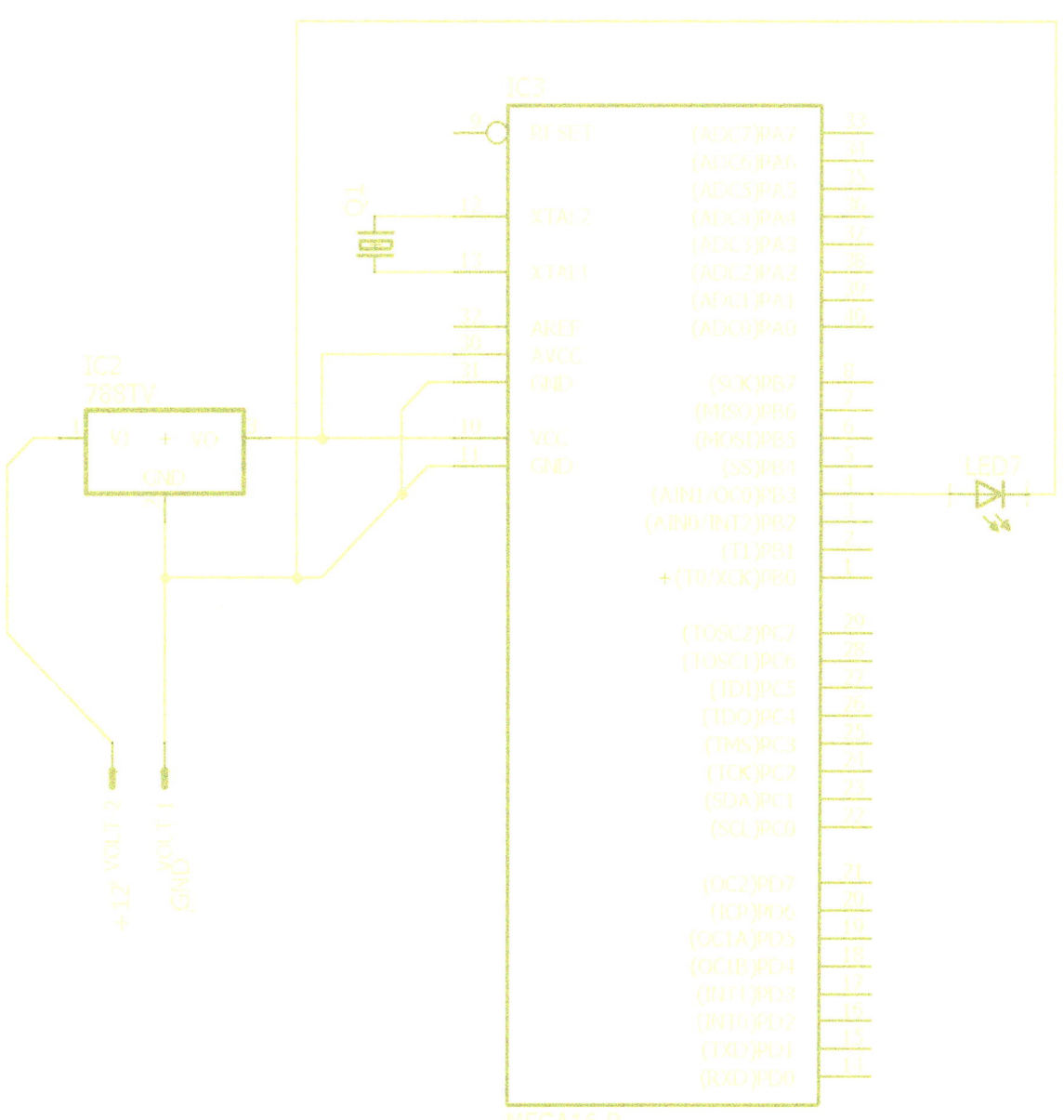

Experiment – 16

LED Controlled using PWM via Button

To control the brightness of an LED you can vary the power which is sent to the LED, for example using a potentiometer (variable resistor), the more power the LED receives the brighter it is, the less power it receives the dimmer it is. Microcontrollers are digital, meaning they only have two 'power' states, on and off. Although it is possible to supply a varying power from a microcontroller (using a Digital to Analogue Convertor (DAC)) this usually requires an additional chip. PWM provides the ability to 'simulate' varying levels of power by oscillating the output from the microcontroller.

If, over a short duration of time, we turn the LED on for 50% and off for 50%, the LED will appear half as bright since the total light output over the time duration is only half as much as 100% on. The important factor here is the 'duration', if we turn the light on and off too slowly the viewer will see the flashing of the LED not a constant light output which appears dimmer. The pulsing width (in this case 50%) is the important factor here. By varying (or 'modulating') the pulsing width we can effectively control the light output from the LED, hence the term PWM or Pulse Width Modulation.

When using PWM it's important to consider how slowly we can 'flash' the LED so that the viewer does not perceive the oscillation. The eye's inability to see rapid oscillations of light is caused by our 'persistence of vision' which means, in very simple terms, we see the light as on even after it has turned off. This technique is how televisions display a seemingly moving picture which is actually made up of a number of different still frames displayed one after the other very rapidly. The minimum speed of an LED oscillating which can be seen by the human eye varies from person to person. However, for the purposes of this article, we will use a minimum speed of 50Hz, or 50 times per second.

When using PWM there are certain terms which you will come across again and again. The most important term is 'duty-cycle'. The duty-cycle refers to the total amount of time a pulse is 'on' over the duration of the cycle, so at 50% brightness the duty-cycle of the LED is 50%. The 'cycle' itself is measured (usually) in Hertz which gives us the cycles-per-second. So at 50Hz our cycle is 1 second divided by 50 cycles, which is 0.02 seconds. Since we are using such small time measurements it's more useful to use microseconds (there are 1,000,000 microseconds in a second), this gives us a cycle duration of 20,000 microseconds which is 50 cycles per second or 50Hz. During the 20,000 microseconds we have to turn the LED either on or off depending on the required duty-cycle so, for example, a 75% duty-cycle requires the pulse to be on for 15,000 microseconds and then off for 5,000 microseconds. To perform the PWM using an interrupt we have to call the interrupt once every 1,000 microseconds and decide if the LED should be on or off. To do this we have to set up a timer on the microcontroller which calls the interrupt when it expires.

In this practice we are using button to increase and decrease the intensity of LED. When button 1 is press the LED will increase its intensity slowly and go to its highest brightness, when button 2 is press the LED will decrease its intensity slowly till it's off state. The smoothness of glow ON and glow OFF is depend on delay we are using. We are interfacing LED on PORT B and switching ON – OFF smoothly using PWM.

Programming is done in "AVRstudio4 " and header are mention in program are mention at the end. The component are used are given below.

WARNING

The header file used "PWM.h " is important header where port pin is predefine for LED. Header is mention at the end of practical in 'required appendix'

Component details

1) AVR ATmega16 IC3

2) 7805 IC2

3) Crystal 16 MHz

4) Push-to-on button

5) LED

6) 9 volt battery cap

7) ISP burner

Program

```c
#include<avr/io.h>
#include<util/delay.h>
#include<nir/multiutil.h>
#include<nir/PWM.h>
void main()
{
DDRC = 0b11111111;
PORTC = 0b00000000;
uint8_t brightness=0;    // Unsigned Integers of 8 bits
InitPWM();               // initialize PWM from header
while(1)
{

if(PINC == 0b00000001)
    {
        for(brightness=0;brightness<255;brightness++) // for loop for increasing LED light
        {
        SetPWMOutput(brightness);
            _delay_ms(250);          // this delay justify the smoothness of glowing up LED
        }


}
if(PINC == 0b00000010)
    {
        for(brightness=255;brightness>0;brightness--)// for loop for decreasing LED light
        {
            SetPWMOutput(brightness);
```

                _delay_ms(250);   // this delay justify the smoothness of glowing down LED
            }
        }
}
}

Schismatic Diagram

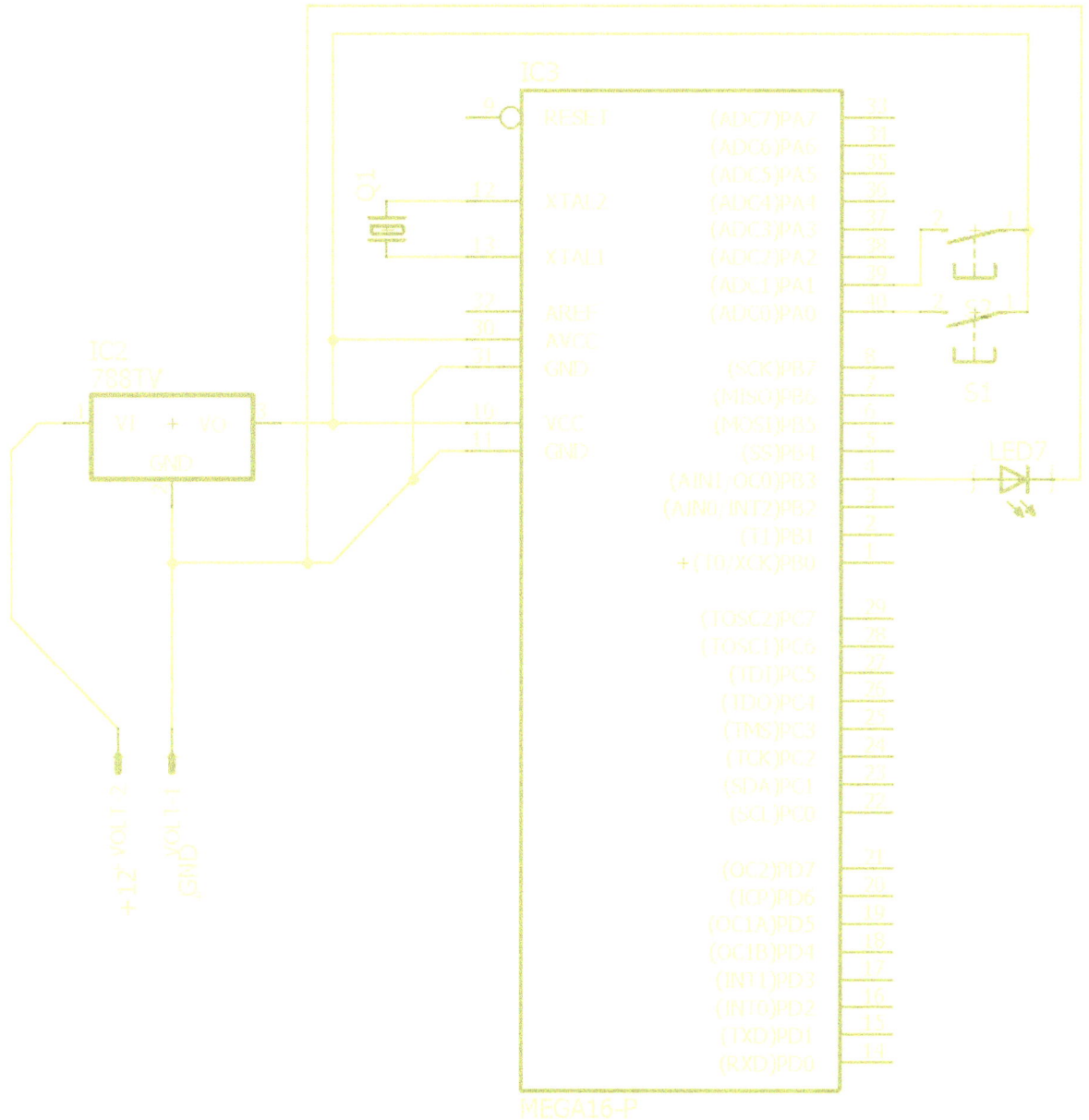

Experiment – 17

LCD Brightness controlling

We are using various types LCD in day-to-day life, Some of them are 16x2, 16x4, 20x5, graphical LCD etc. There is one parameter to be adjusted for contrast of letter to be display at LCD pin Vss. This pin of LCD is connected to POT to control the contrast of letter. There are 2 more pin "led+ & led-" this pin controllers backlight of LCD where we can see digits in dark also. In this practice we are changing led point with LCD LED point to control backlight of LCD.

To control the brightness of an LED you can vary the power which is sent to the LED, for example using a potentiometer (variable resistor), the more power the LED receives the brighter it is, the less power it receives the dimmer it is. Microcontroller's are digital, meaning they only have two 'power' states, on and off. Although it is possible to supply a varying power from a microcontroller (using a Digital to Analogue Convertor (DAC)) this usually requires an additional chip. PWM provides the ability to 'simulate' varying levels of power by oscillating the output from the microcontroller.

There are wide range of conditions over which LCD are used means that it is desirable to produce displays whose brightness can be altered to match both bright and dim environments. This allows a user to set the screen to a comfortable level of brightness depending on their working condition. maker will normally quote a maximum brightness figure in their display specification, but it is also important to consider the lower range of adjustments possible from the screen as you would probably never want to use it at its highest setting. If you will certainly need to use the screen at something a little less harsh on the eyes. As a reminder, we test the full range of backlight adjustments and the corresponding brightness values during each of our reviews. During our calibration process as well we try to adjust the screen which is considered the recommended brightness for an LCD in normal lighting conditions. This process helps to give you an idea of what adjustments you need to make to the screen in order to return a brightness which you might actually want to use day to day.

Changing the display brightness is achieved by reducing the total light output for both CCFL- and LED-based backlights. By far the most prevalent technique for dimming the backlight is called Pulse Width Modulation (PWM), which has been in use for many years in desktop and laptop displays. However, this technique is not without some issues and the introduction of displays with high brightness levels and the popularisation of LED backlights has made the side-effects of PWM more visible than before, and in some cases may be a source of visible flicker, eyestrain, eye fatigue, headaches and other associated issues for people sensitive to it. This article is not intended to alarm, but is intended to show how PWM works and why it is used, as well as how to test a display to see its effects more clearly. We will also take a look at the methods some manufacturers are now adopting to address these concerns and provide flicker-free backlights instead. As awareness grows, more and more manufacturers are focusing on eye health with their monitor ranges.

Component details

1) AVR ATmega16 IC3
2) 7805 IC2
3) Crystal 16 MHz
4) Push-to-on button
5) 16x2 LCD
6) 9 volt battery cap
7) ISP burner

Program

```c
#include<avr/io.h>
#include<util/delay.h>
#include<nir/multiutil.h>
#include<nir/PWM.h>
void main()
{
DDRD = 0b11111111;
PORTD = 0b00000000;
uint8_t brightness=0;   // Unsigned Integers of 8 bits
InitPWM();              // initialize PWM from header
while(1)
{

                lcd_clrscr();
                lcd_goto(1,1);
                lcd_prints("LCD BTR CTR");
                _delay_ms(100);


if(PIND == 0b00000001)
        {
                for(brightness=0;brightness<255;brightness++) // for loop for inc LED light
                {
                        SetPWMOutput(brightness);
                        _delay_ms(250);    // this delay justify the smoothness of glowing up
                                           LED

                }
```

```
}
if(PIND == 0b00000010)
    {
        for(brightness=255;brightness>0;brightness--)
        {
            SetPWMOutput(brightness);
            _delay_ms(250);

        }
    }
}
}
```

Schematic Diagram

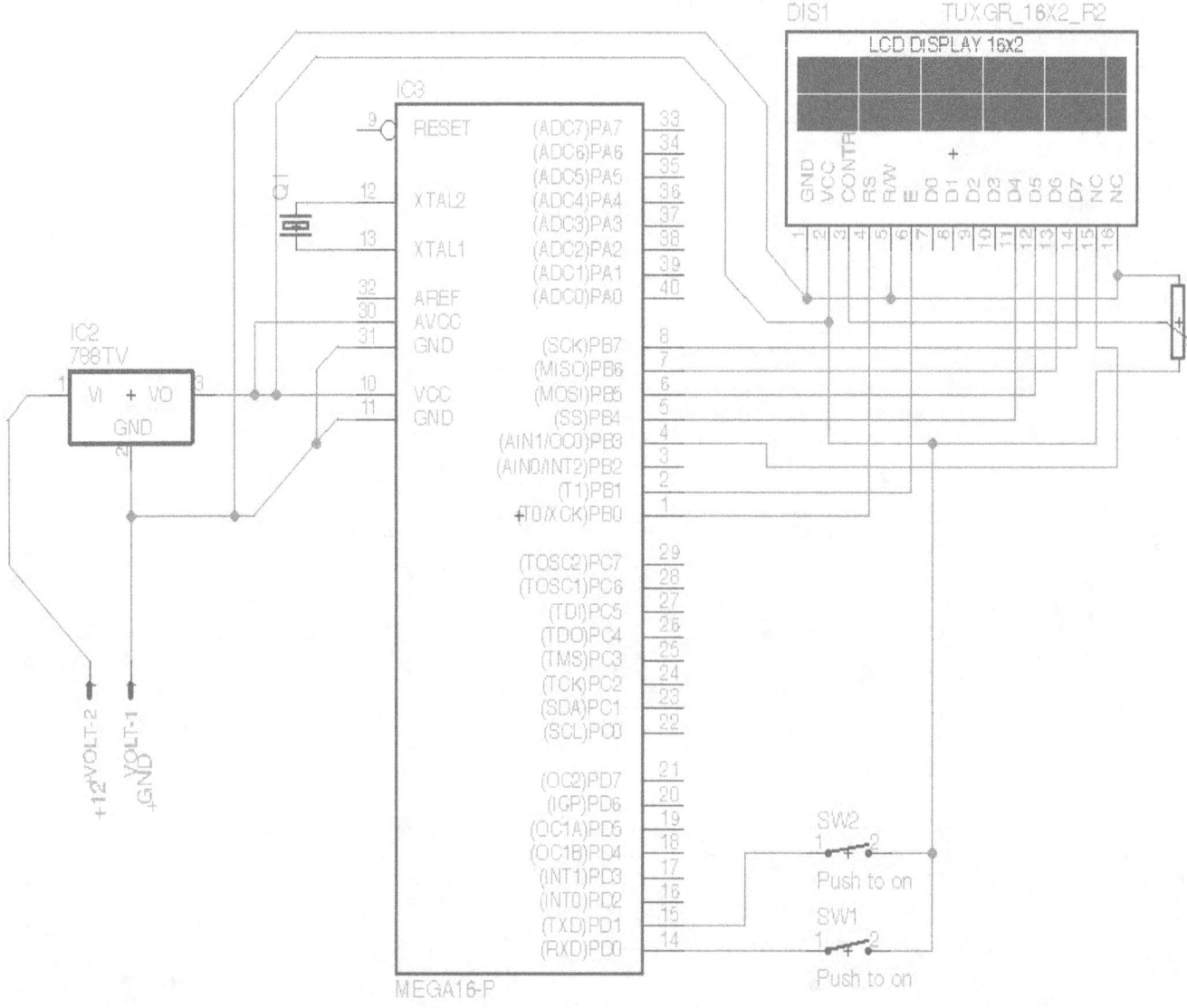

Experiment – 18

DC motor speed controlling using PWM

In early practices we are going to control the speed of motor using PWM technique. The concept behind controlling speed of DC motor is converting the nature of signal from digital to analog using pulse width modulation. The main application of PWM uses digital signals to control power applications, as well as being fairly easy to convert back to analog with a minimum of hardware.

Analog systems, such as linear power supplies, tend to generate a lot of heat since they are basically variable resistors carrying a lot of current. Digital systems don't generally generate as much heat. Almost all the heat generated by a switching device is during the transition (which is done quickly), while the device is neither on nor off, but in between.

PWM can have many of the characteristics of an analog control system, in that the digital signal can be freewheeling. PWM does not have to capture data, although there are exceptions to this with higher end controllers.

One of the parameters of any square wave is duty cycle. Most square waves are 50%, this is the norm when discussing them, but they don't have to be symmetrical. The ON time can be varied completely between signal being off to being fully on, 0% to 100%, and all ranges between.

Shown below are examples of a 10%, 50%, and 90% duty cycle. While the frequency is the same for each, this is not a requirement.

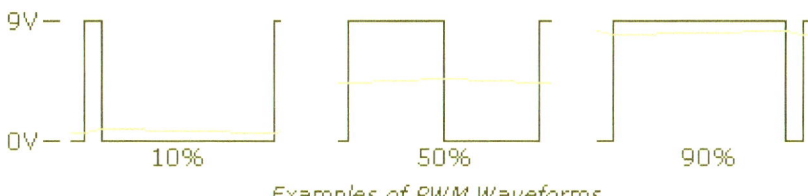

Examples of PWM Waveforms

The reason PWM is popular is simple. Many loads, such as resistors, integrate the power into a number matching the percentage. Conversion into its analog equivalent value is straightforward. LEDs are very nonlinear in their response to current, as we have seen in last practical the parameter of LED using PWM which give an LED half its rated current still get more than half the light the LED can produce. With PWM the light level produced by the LED is very linear. Motors, which will be covered, are also very responsive to PWM.

As we are going to practice application is motor speed control. Motors as a class require very high currents to operate. Being able to vary their speed with PWM increases the efficiency of the total system by quite a bit. PWM is more effective at controlling motor speeds at low RPM than linear methods.

PWM is often used on L293D on enable pin to control the speed of motor, for high current rating devices we can also use MOSFET. In this practice we are only using the motor driver L293D to perform the speed control of the DC motor

WARNING

The header file used "PWM.h " is important header where port pin is predefine for LED. Header is mention at the end of practical in 'required appendix'

This practical is only use for 12v@500ma DC purpose.

Component details

1) AVR ATmega16 IC3
2) 7805 IC2
3) Crystal 16 MHz
4) Push-to-on button
5) L293D
6) 12v DC Motor
7) 12 volt battery adaptor
8) ISP burner

Program

#include<avr/io.h>

#include<util/delay.h>

```c
#include<nit/PWM.h>
void main()
{
DDRD = 0b11111111;
PORTD = 0b00000000;
DDRA = 0b11111111;
PORTA = 0b00000000;
uint8_t brightness=0;
InitPWM();
while(1)
{



if(PINA == 0b00000001)
    {


        PORTD = 0b00000010
        _delay_ms(50);
        for(brightness=0;brightness<255;brightness++)
        {
            SetPWMOutput(brightness);
            _delay_ms(250);


        }

}
if(PIND == 0b00000010)
    {
        PORTD = 0b00000001;
```

```
        _delay_ms(50);

        for(brightness=255;brightness>0;brightness--)// for loop for decreasing LED light

        {

            SetPWMOutput(brightness);

            _delay_ms(250);                    // this delay justify the smoothness of
                                                glowing down LED


        }
    }
}
}
```

Schematic Diagram

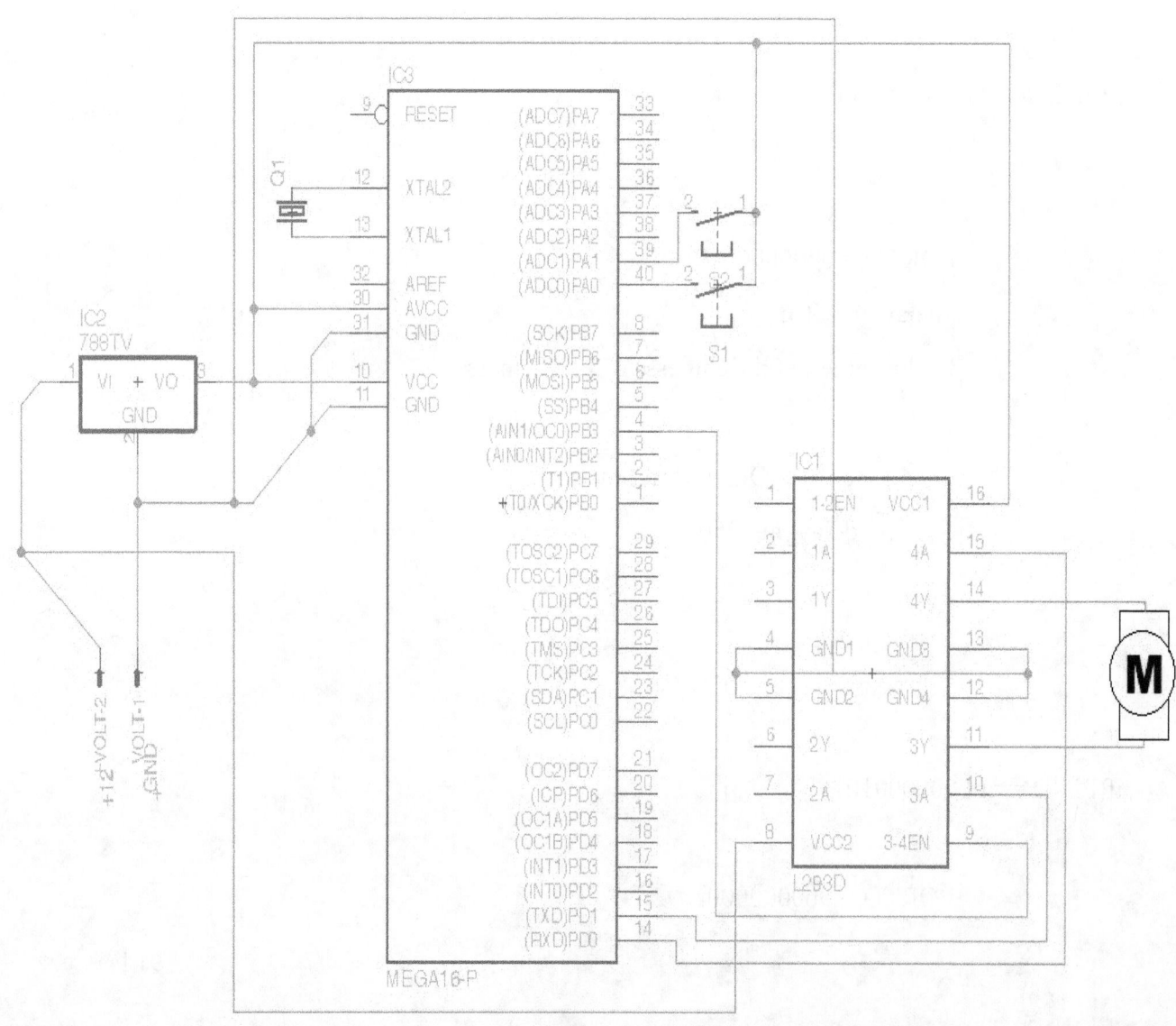

Experiment – 19

Voltage Step-down Without Transformer

In this practical low voltage DC power supply from AC power is extracting without transformer. The most important problem faced in electronics circuits are weighted and bulky size of transformer. The straight forward technique is the use of a step down transformer to reduce the 230 V or 110V AC to a preferred level of low voltage AC. We are going to do practice on constructing power supply for your development without a transformer. This circuit is so simple and it uses a voltage dropping capacitor in series with the phase line. Transformer less power supply is also called as capacitor power supply. It can generate 5V, 6V, 12V 150mA from 230V or 110V AC by using appropriate zener diodes.

This transformer less power supply circuit is also named as capacitor power supply since it uses a special type of AC capacitor in series with the main power line. A common capacitor will not do the work because the mains spikes will generate holes in the dielectric and the capacitor will be cracked by passing of current from the mains through the capacitor.

X rated capacitor suitable for the use in AC mains is vital for reducing AC voltage. A X rated dropping capacitor is intended for 250V, 400V, 600V AC. Higher voltage versions are also obtainable. The dropping capacitor is non-polarized so that it can be connected any way in the circuit.

The 470kΩ resistor is a bleeder resistor that removes the stored current from the capacitor when the circuit is unplugged. It avoid the possibility of electric shock. Reduced AC voltage is rectified by bridge rectifier circuit. We have already discussed about bridge rectifiers. Then the ripples are removed by the 1000μF capacitor.

This circuit provides 24 volts at 160 mA current at the output. This 24 volt DC can be regulated to necessary output voltage using an appropriate 1 watt or above zener diode. Here we are using 6.2V zener. You can use any type of zener diode in order to get the required output voltage.

Design of System

Reactance of the capacitor

$$X = 1/2\pi f C$$

where f is the supply frequency and C is the capacitance. If the supply frequency is 50Hz, then reactance of 2.2µF X rated capacitor is given by,

$$X = \frac{1}{2\pi * 50 * 2.2 * 10^{-6}}$$

$$= 1.44 k\Omega$$

So output current,

$$I = V/X$$

$$= 230/1.44 * 10^3$$

$$= 159 mA$$

Tolerance +/- 15%

WARNING

Do not touch at any points in the circuit because some points are at mains potential. Keep away from touching the points around the dropping capacitor to prevent electric shock, even after switching off the mains. Great care should be taken to construct the circuit since there isno isolation between mains and our circuit. Adequate spacing must be given between the components.

Schematic Diagram

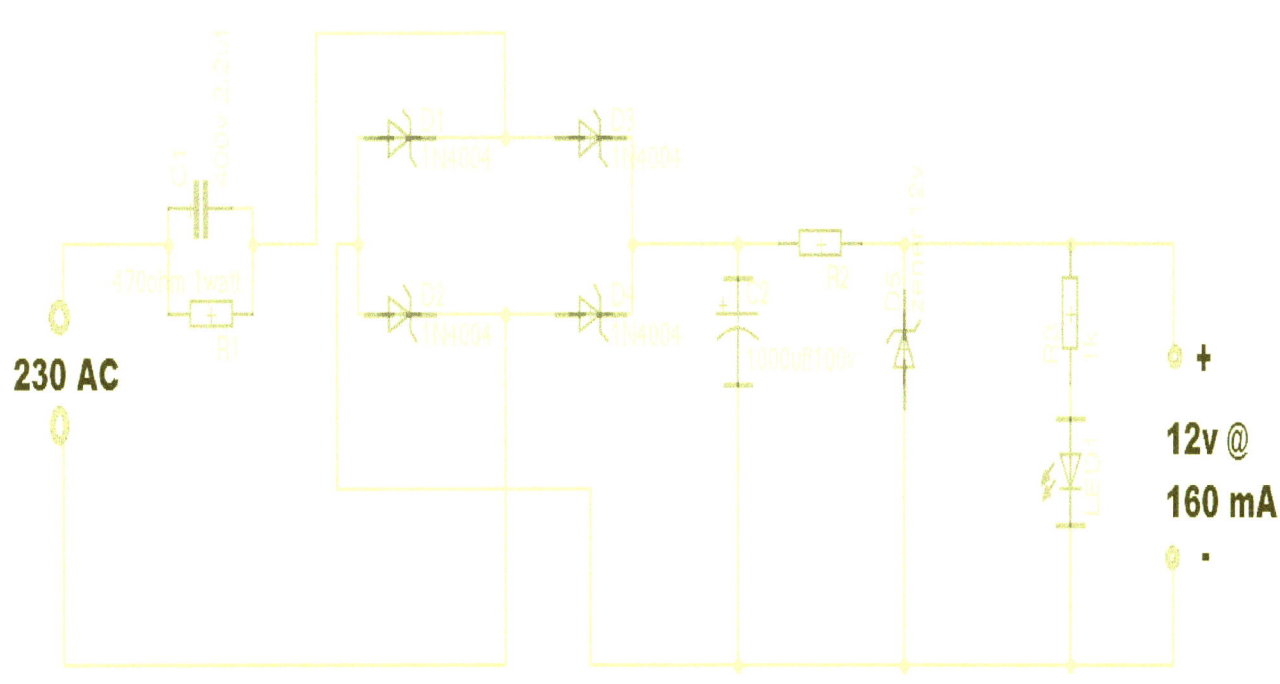

Experiment – 20

LASER Security

In this practical we are going to study main components of the laser security system. It works based on LASER detection from LASER transmission to LDR in case of any security fault. Based on this alarm unit is triggered. The system alerts the security monitoring person and the local law enforcement body if required. At the same time a high pitched sound also sirens. There are two types of laser beams are available color for ex green blue red laser, and infrared modules. System via an infrared module will be not visible. Another beam like color laser would be visible to the naked eye and serve as a deterrent purpose.

Laser security alarm is based on the interruption of Laser beam. The laser pointer is used as the source of light beam. If somebody tries to break the laser path, then an alarm will be triggered. Normally laser door alarm circuit will have two sections. First one, laser transmitter is a laser pointer readily available. This is powered with 3 volt DC supply and fixed on one side of the door frame. The receiver will have a LDR at the front end.

In this practices we are using ATmega16 as we require ADC to measure LDR reading. We studied the working and practice the ADC and LDR.

WARNING

LASER Diode works on 3.3 volt and it and easily damage naked eye.

Note:-

The command line written in blue color like – "lcd_clrscr();" are temporary program line to understand the analog value given out from sensor, once you calibrate the analog value for sensor note this value in program for condition such as at my end for forward direction x was triggering at greater than 410 so I have put it in condition "(x > 680)". When you wll calibrate all analog value **you can delete this line or can make comment to it**.

Component details

1) AVR ATmega16 IC3
2) 7805 IC2
3) Crystal 16 MHz
4) LASER Diode
5) LDR
6) Buzzer
7) 12 volt battery adaptor
8) ISP burner

Program

#include<avr/io.h>

#include<nir/multiutil.h>

#include<util/delay.h>

main()
{
    int a;
    init_adc();
    lcd_init();

    DDRC=0b00000000;

```c
PORTC=0b11111111;

while(1)
    {
        a=read_adc(0);
        _delay_ms(50);

        lcd_clrscr();
        lcd_goto(1,1);
        lcd_printi(x);
        _delay_ms(150);



        if(a< 995)
        {
        PORTC=0b00000000;
        _delay_ms(50);
        lcd_clrscr();
        lcd_goto(1,2);
        lcd_prints("warning");



        }
        else
        {
        PORTC=0b11111111;
        _delay_ms(5550);
        lcd_clrscr();
        lcd_goto(1,2);
        lcd_prints("Laser Security");
```

}

}
}

Schematic Diagram

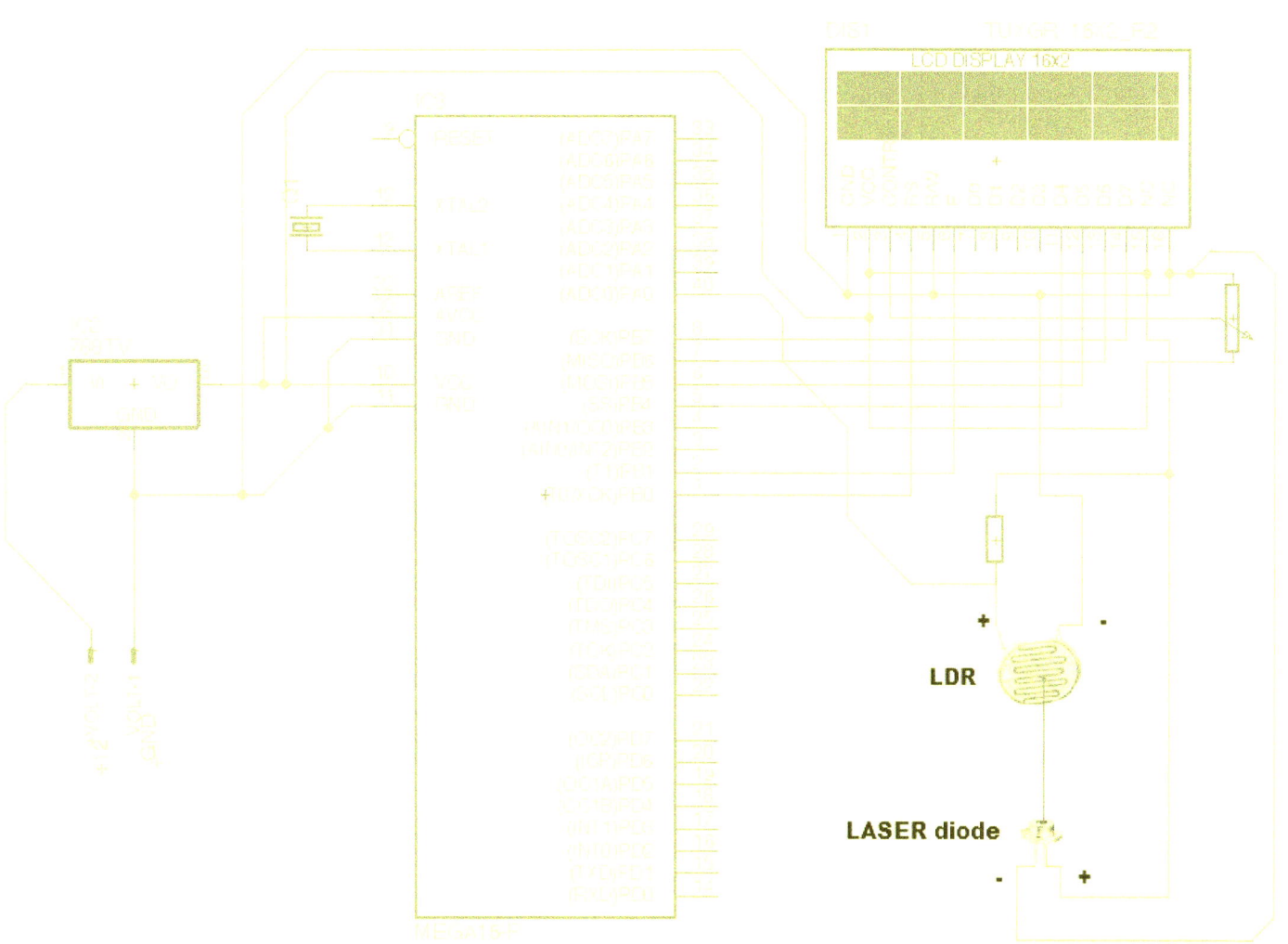

Experiment – 21

Temperature sensor interfacing

This practical we are using IC LM35 as a sensor for detecting accurate centigrade temperature. Linearity defines how well over a range of temperature a sensor's output consistently changes. Unlike thermistor, Linearity of a precision IC Sensors are very good of 0.5°C accuracy and has wide temperature range. Its output voltage is linearly proportional to the Celsius (Centigrade) temperature.

The LM35 is rated to operate over a -55° to +150°C temperature range. It draws only 60 µA from its supply, it has very low self-heating, less than 0.1°C in still air. LM35 Operates from 4 to 30 volts.

Output of IC is 10mv/degree centigrade for eg if the output of sensor is 280 mV then temperature is 28 degree C. so by using a Digital multimeter we can easily calculate the degree temperature.

LM35 by National Semiconductor is a popular and low cost temperature sensor. To use the sensor simply connect the Vcc to 5V, GND to Ground and the Out to one of the ADC (analog to digital converter channel). The output linearly varies with temperature. The analog output of LM35 can be readied as 10 MilliVolts per degree centigrade. So if the output is 310 mV then temperature is 31 degree C. To perform this practical we have already studied and practice the ADC of AVRs and also using 16x2 LCD display

In this circuit we are using Atmega16 as we are working on analog reading and the inbuilt ADC of AVR is used to convert the analog voltage from the LM35 to digital value.

The resolution of AVRs ADC is 10bit and for reference voltage we are using 5V so the resolution in terms of voltage is

5/1024 = 5mV approximately

So if ADC's result corresponds to 5mV i.e. if ADC reading is 10 it means

10 x 5mV = 50mV

You can get read the value of any ADC channel using the function

"read_adc(x);" where "x" indicates the channel of adc.

Component details

1) AVR ATmega16 IC3

2) 7805 IC2

3) Crystal 16 MHz

4) LM35

5) Buzzer

6) 12 volt battery adaptor

7) ISP burner

Program

```c
#include <avr/io.h>
#include<util/delay.h>
#include<swits.h>
main()
{

    init_adc();
    lcd_init();

    unsigned int val;
    unsigned int voltage;
    unsigned int ts;


while(1)
{
            val=read_adc(1);
            ts=(((val)/1023.0)*5*100);  //calculation to convert volt to degree
```

```c
            ts=round(val*0.48876
            lcd_goto(1,1);
            lcd_printi(ts);
            lcd_goto(1,1);
            lcd_prints("D");
            _delay_ms(350);



    if( ts > 50)
            {
            lcd_clrscr();
            lcd_goto(1,2);
            lcd_prints("warning");
            _delay_ms(350);
            lcd_goto(1,2);
            lcd_prints("Temp Excid");
            _delay_ms(350);
            }
        }
}
```

Schematic Diagram

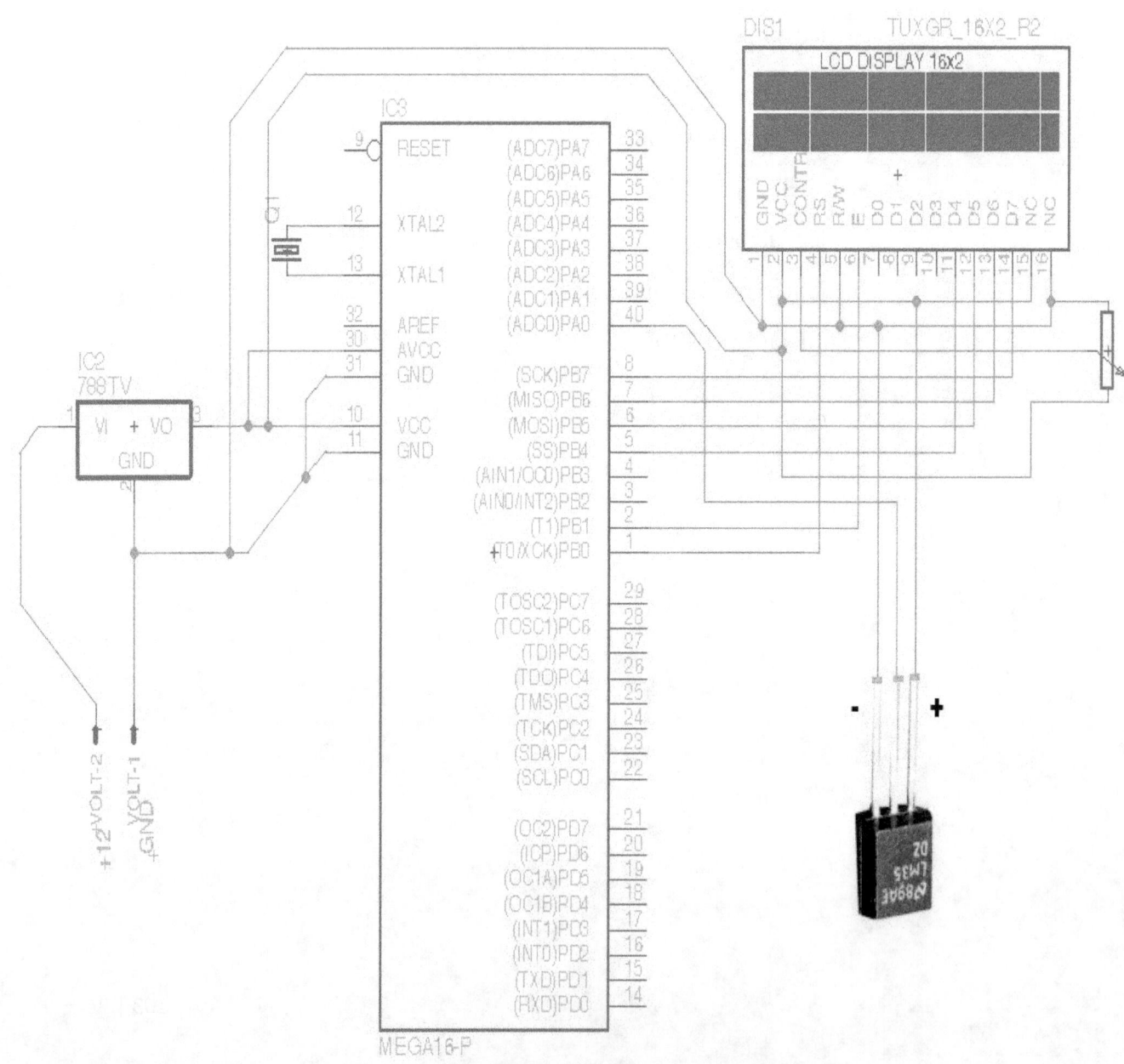

Experiment – 22

PIR sensor interfacing

In this practical we are going to practice the detection of motion of any object passing from facing side of sensor. We are using PIR sensor to detect the motion, it stands for Passive Infrared. The basic working operation to detect the motion of PIR as all objects with a temperature above absolute zero emit heat energy in the form of radiation. Usually this radiation is invisible to the human eye because it radiates at infrared wavelengths, but it can be detected by electronic devices designed for such a purpose.

The term *passive* in this instance refers to the fact that PIR devices do not generate or radiate any energy for detection purposes. They work entirely by detecting the energy given off by other objects. PIR sensors don't detect or measure "heat"; instead they detect the infrared radiation emitted or reflected from an object.

A PIR-based motion detector is used to sense movement of people, animals, or other objects. They are commonly used in burglar alarms and automatically-activated lighting systems. They are commonly called simply "PIR", or sometimes "PID", for "passive infrared detector".

PIR sensor is the abbreviation of Passive Infrared Sensor. It measures the amount of infrared energy radiated by objects in front of it. They does not emit any kind of radiation but senses the infrared waves emitted or reflected by objects. The heart of a PIR sensor is a solid state sensor or an array of such sensors constructed from pyro-electric materials. Pyro-electric material is material by virtue of it generates energy when exposed to radiation. Gallium Nitride is the most common material used for constructing PIR sensors. Suitable lenses are mounted at the front of the sensor to focus the incoming radiation to the sensor face. When ever an object or a human passes across the sensor the intensity of the of the incoming radiation with respect to the background increases. As a result the energy generated by the sensor also increases. Suitable signal conditioning circuits convert the energy generated by the sensor to a suitable voltage output. In simple words the output of a PIR sensor module will be HIGH when there is motion in its field of view and the output will be LOW when there is no motion.

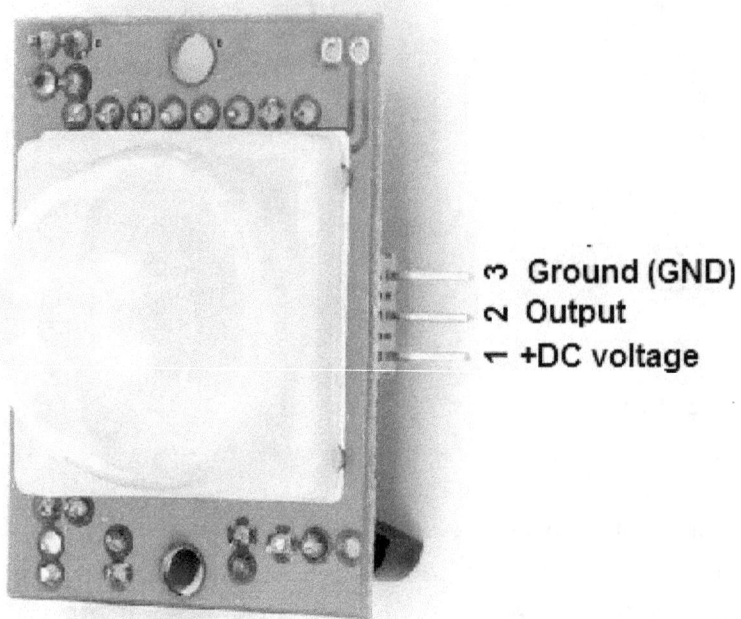

DSN-FIR800 is the PIR sensor module used in this project. It operates from 4.5 to 5V supply and the stand by current is less than 60uA. The output voltage will be 3.3V when the motion is detected and 0V when there is no motion. The sensing angle cone is 110° and the sensing range is 7 meters. The default delay time is 5 seconds. There are two preset resistor on the sensor module. One is used for adjusting the delay time and the other is used for adjusting the sensitivity. Refer the datasheet of DSN-FIR800 for knowing more.

Note:-

The command line written in blue color like – "lcd_clrscr();" are temporary program line to understand the analog value given out from sensor, once you calibrate the analog value for sensor note this value in program for condition such as at my end for forward direction x was triggering at greater than 410 so I have put it in condition "(x > 570)". When you wll calibrate all analog value **you can delete this line or can make comment to it**.

Component details

1) AVR ATmega16 IC3
2) 7805 IC2
3) Crystal 16 MHz
4) PIR sensor

5) Buzzer

6) 12 volt battery adaptor

7) ISP burner

Program

```c
#include<avr/io.h>
#include<swits.h>
#include<util/delay.h>
main()
{
    int x;
    init_adc();
    lcd_init();

    while(1)
    {
        x=read_adc(0);

        _delay_ms(150);
        lcd_clrscr();
        lcd_goto(1,1);
        lcd_printi(x);
        _delay_ms(150);


        if(x > 570)
        {
        lcd_clrscr();
        lcd_goto(1,1);
        lcd_prints("Motion Detected");
        _delay_ms(5000);

        }
    else
        {
        lcd_clrscr();
        lcd_goto(1,1);
        lcd_prints("Motion Sensor");
        _delay_ms(150);

        }


    }
}
```

Schematic Diagram

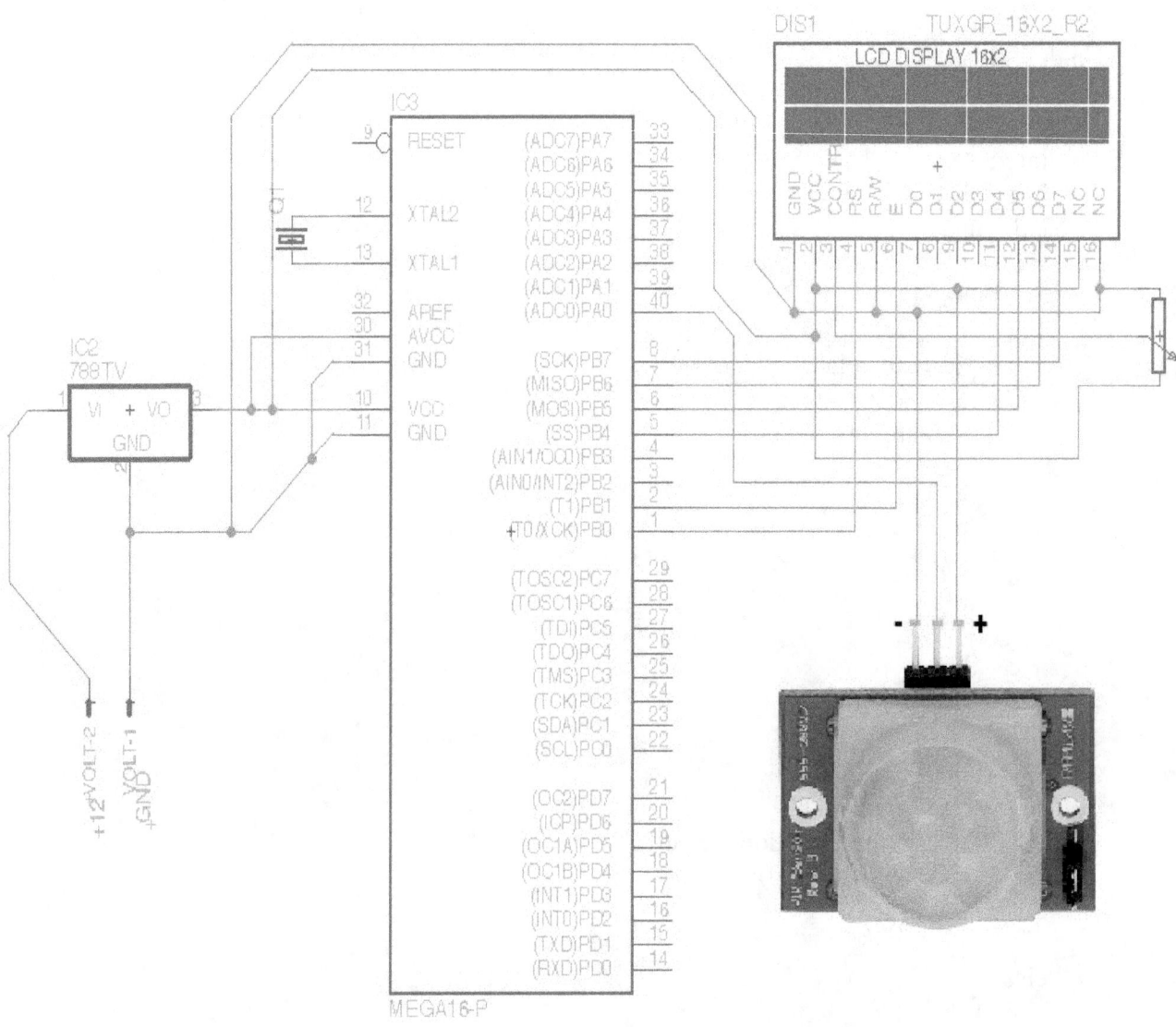

Experiment – 23

4x4x4 LED Cube

In this practice we are going to perform the simple concept of LED ON-OFF to get 3d view and depth of the object. The cube we are making are of 4x4x4 LED which consist of 64 LED (color of any choice), we can also practice more combination such as 2x2x2, 4x4x4, 8x8x8, 16x16x16. As the number of LED increase the hardware complicity also increases

The blinking can be perform on variety of controller but we are working on ATmega8, ATmega16 and Atmega32 according to program flash required.

LED cubes rely on an optical phenomenon called persistence of vision or POV. If you flash an LED really fast, the image stays on your retina for a little while after the LED turns off. By flashing each layer of the cube one after another really fast, it gives the illusion of a 3D image, this is also called multiplexing. The LED cube is made up of columns and layers. Each of the 16 (anode) columns and the four (cathode) layers are connected to the controller board with a separate wire and can be controlled individually.

Construction of the cube

The construction of LED cube is quite simple and easy to grasp.

We have to construct four layers/rows of 16-LEDs in each. These four rows will then be connected together. It is clear that we have 20-control lines which will be responsible for the LED ON/OFF.

Out of these 20, four lines i.e. Rows (R1,R2,R3,R4) are anodes(+).

The columns are Cathodes(-).

Working: To understand the working, let us consider that we want to ON an LED at position (C0,R1). For this we send a HIGH at R1 Terminal of cube through a transistor (A transistor is used because current through microcontroller is not sufficient to drive more than 1-2 LEDs) and a LOW at C0 terminal. This makes a complete path for the current and LED glows.

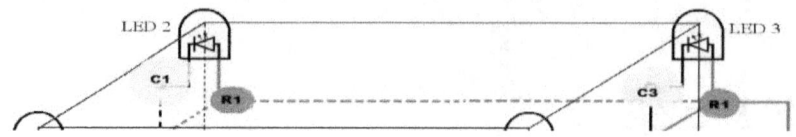

Follow some step to construct LED cube

Soldering grids of 4x4 LEDs freehand would look terrible. To get even-looking 4x4 LED grids, we'll use a template to hold them in place. Also, to minimize adding or cutting wire, we'll use the LED legs to connect the LEDs together.

Find a piece of pegboard that already has a 4x4 grid pre-drilled with 1" spacing between holes. Double fold a piece of aluminum foil over the board and tape it down. The foil will hold the

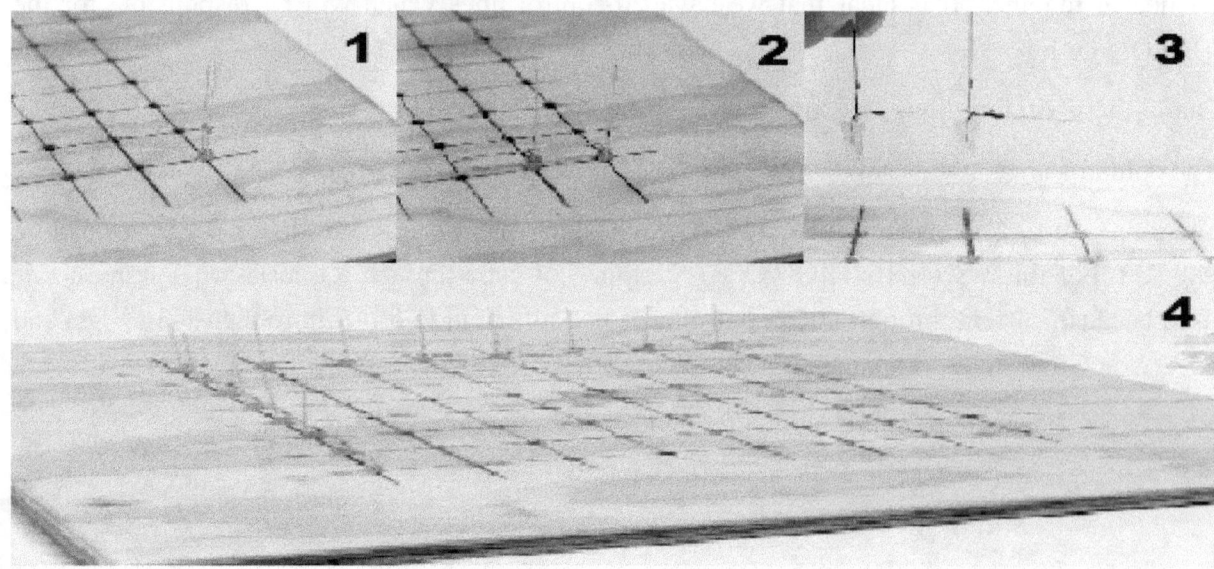

LEDs in place and protect the board while soldering. Use one LED to punch an LED-sized hole through the foil for each hole.

Complicated way :- Find a piece of wood large enough to make a 1" 4x4 grid (leave a little extra room). Draw up a 4x4 grid of lines with a spacing of 1". Make dents at the intersect points with a center punch. Drill 16 holes small enough so that the LED will stay firmly in place and big enough so that the LED can easily be pulled out (without bending the wires).

It is vital to have functioning LEDs, the easier to test the individual LEDs before you solder them together test them individually. Sticking an iron to desolder a damaged LED in the middle of your cube. Take the time to test them.

You can hook the LEDs up to a 3 volt power supply and briefly powering on, use an LED Tester, or simply use a coin cell battery. Hold the coin cell between the legs of the LED and then squeeze the legs. You don't need a resistor since the coin cell runs at 3V and you are only touching it for couple seconds.

To make the cube's four layers of 4x4 LEDs, bend their cathodes (the shorter lead) and solder them together. You get to learn from my mistakes. Here are some soldering tips.

Soldering iron hygiene: Keep the soldering iron clean. That means wiping it on the sponge every time you use it or whenever you see the tip becoming dirty with flux or oxidization, even if you are in the middle of soldering. Having a clean soldering tip makes it a lot easier to transfer heat to the soldering target.

Soldering speed: Get in and out quickly. Apply a tiny amount of solder to the iron tip. Touch the part you want to solder with the side of your iron where you just put a little solder. Let the target heat up for 0.5-1 seconds and then touch the other side of the target you are soldering with the solder. Remove the soldering iron immediately after applying the solder.

Mistakes and cool down: If you make a mistake (for example if the wires move before the solder hardens or you don't apply enough solder), *do not try again right away*. The LED is already very hot and applying more heat with the soldering iron will only make it hotter.

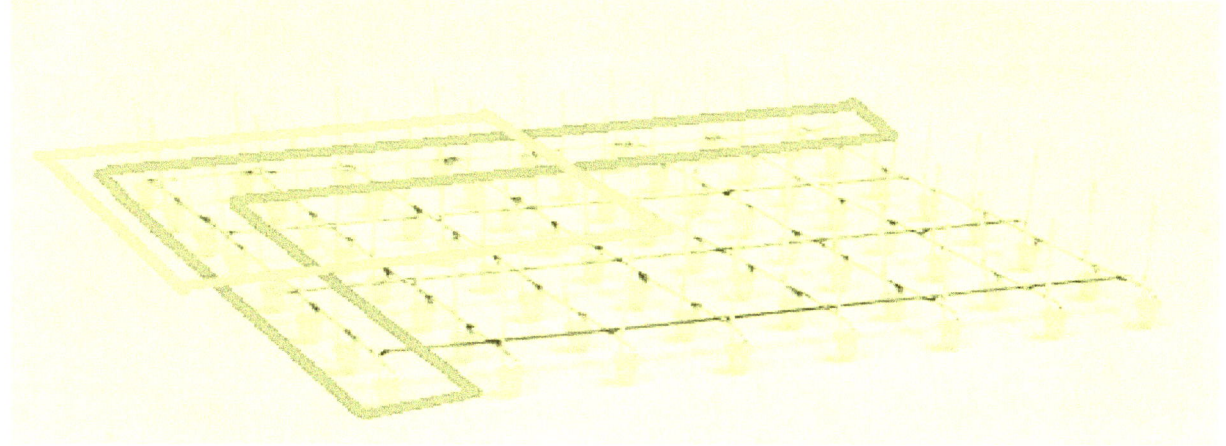

Continue with the next LED and let it cool down for a minute or blow on it to remove some heat.

To create your layer, place LEDs in template on two outer rows in an "L" shape and solder them together. Continue to insert LEDs row by row and soldering them together (going one row at a time leaves you space to solder) until you complete the rows. Add wire cross bracing in the front where the led rows are not connected (use the same hookup wire but strip the plastic coating off and straighten the wire).

Solder the Layers Together

Take your time while building the layers. The quality and look of your final cube depends on the layers be built neatly and evenly. Select your best layer piece and put it back in the template. This will be the top layer. It can get tricky holding one layer above the other while soldering. You can use a third hand or used a 9V battery, which is the perfect size to help create the correct spacing.

Warning: Tape over the battery poles to avoid accidentally overloading the LEDs while soldering.

Carefully align the layers and solder the corner LEDs. Next, solder all the LEDs around the edge of the cube, moving the 9V batteries along as you go around ensuring that the layers are soldered in parallel. Then move a 9V battery to the middle of the cube, sliding it in from one of the sides, and solder a couple of the LEDs in the middle.

Your cube should be fairly stable now, so you can continue soldering the rest of the LEDs without using any extra support. When you have soldered all the columns, it is time to test the LEDs again. Remember that tab sticking out from the upper right corner of the layer Now it's time to use it to test your LEDS.

Simple idea to test cube by fixing two wires on a 3V battery with tape and using that to test. Touch the negative wire to the layer we want to test and touch the positive wire to the column you want to test and should see an individual LED light up. Continue touching each column in each layer to be sure they are all functioning.

Calculation of the pins required for cubes:
    For 4x4x4 cube
        Rows-4 pins

Columns- 4x4=16 pins

{ For 8x8x8 cube

Rows-8 pins

Columns- 8x8=64 pins

(However 64 pins are not there in low profile uC, so this can be done through Shift Registers)

Similarly 16x16x16 and further can be constructed.

Warning

An important thing which must be kept in mind must be the current consumption by the LEDs.
Proper supply should be provided
High current capacity cube can be design properly otherwise it can damage controller easily.

Component details

1) AVR ATmega16 IC3

2) 7805 IC2

3) Crystal 16 MHz

4) LED (any color 64)

5) Transistor BC547

6) 12 volt battery adaptor

7) ISP burner

Program

```c
#include<avr/io.h>
#include<util/delay.h>

void main()
{
DDRA = 0b00000000;
PORTA = 0b11111111;
DDRB = 0b00000000;
PORTB = 0b11111111;
DDRD = 0b11111111;
PORTD = 0b00000000;

while(1)
{

            _delay_ms(5);
            PORTA = 0b11111110;
            PORTB = 0b11111111;
            PORTD = 0b00000001;     // LED 1
            _delay_ms(5);
            PORTA = 0b11111101;
            PORTB = 0b11111111;
            PORTD = 0b00000001;     // LED 2
            _delay_ms(5);
            PORTA = 0b11111011;
            PORTB = 0b11111111;
            PORTD = 0b00000001;     // LED 3
```

```
_delay_ms(5);
PORTA = 0b11110111;
PORTB = 0b11111111;
PORTD = 0b00000001;
_delay_ms(5);
PORTA = 0b11101111;
PORTB = 0b11111111;
PORTD = 0b00000001;
_delay_ms(5);
PORTA = 0b11011111;
PORTB = 0b11111111;
PORTD = 0b00000001;
_delay_ms(5);
PORTA = 0b10111111;
PORTB = 0b11111111;
PORTD = 0b00000001;
_delay_ms(5);
PORTA = 0b01111111;
PORTB = 0b11111111;
PORTD = 0b00000001;
_delay_ms(5);
PORTA = 0b11111111;
PORTB = 0b11111110;
PORTD = 0b00000001;
_delay_ms(5);
PORTA = 0b11111111;
PORTB = 0b11111101;
PORTD = 0b00000001;
_delay_ms(5);
```

```c
PORTA = 0b11111111;
PORTB = 0b11111011;
PORTD = 0b00000001;     // LED 11
_delay_ms(5);
PORTA = 0b11111111;
PORTB = 0b11110111;
PORTD = 0b00000001;     // LED 12
_delay_ms(5);
PORTA = 0b11111111;
PORTB = 0b11101111;
PORTD = 0b00000001;     // LED 13
_delay_ms(5);
PORTA = 0b11111111;
PORTB = 0b11011111;
PORTD = 0b00000001;     // LED 14
_delay_ms(5);
PORTA = 0b11111111;
PORTB = 0b10111111;
PORTD = 0b00000001;     // LED 15
_delay_ms(5);
PORTA = 0b11111111;
PORTB = 0b01111111;
PORTD = 0b00000001;     // LED 16

_delay_ms(5);
PORTA = 0b11111110;
PORTB = 0b11111111;
PORTD = 0b00000010;     //LED 17
_delay_ms(5);
```

PORTA = 0b11111101;
PORTB = 0b11111111;
PORTD = 0b00000010;   // LED 18
_delay_ms(5);
PORTA = 0b11111011;
PORTB = 0b11111111;
PORTD = 0b00000010;   // LED 19
_delay_ms(5);
PORTA = 0b11110111;
PORTB = 0b11111111;
PORTD = 0b00000010;   // LED 20
_delay_ms(5);
PORTA = 0b11101111;
PORTB = 0b11111111;
PORTD = 0b00000010;   // LED 21
_delay_ms(5);
PORTA = 0b11011111;
PORTB = 0b11111111;
PORTD = 0b00000010;   // LED 22
_delay_ms(5);
PORTA = 0b10111111;
PORTB = 0b11111111;
PORTD = 0b00000010;   // LED 23
_delay_ms(5);
PORTA = 0b01111111;
PORTB = 0b11111111;
PORTD = 0b00000010;   // LED 24
_delay_ms(5);
PORTA = 0b11111111;

```c
PORTB = 0b11111110;
PORTD = 0b00000010;     // LED 25
_delay_ms(5);
PORTA = 0b11111111;
PORTB = 0b11111101;
PORTD = 0b00000010;     // LED 26
_delay_ms(5);
PORTA = 0b11111111;
PORTB = 0b11111011;
PORTD = 0b00000010;     // LED 27
_delay_ms(5);
PORTA = 0b11111111;
PORTB = 0b11110111;
PORTD = 0b00000010;     // LED 28
_delay_ms(5);
PORTA = 0b11111111;
PORTB = 0b11101111;
PORTD = 0b00000010;     // LED 29
_delay_ms(5);
PORTA = 0b11111111;
PORTB = 0b11011111;
PORTD = 0b00000010;     // LED 30
_delay_ms(5);
PORTA = 0b11111111;
PORTB = 0b10111111;
PORTD = 0b00000010;     // LED 31
_delay_ms(5);
PORTA = 0b11111111;
PORTB = 0b01111111;
```

```
PORTD = 0b00000010;        // LED 32
_delay_ms(5);
PORTA = 0b11111110;
PORTB = 0b11111111;
PORTD = 0b00000100;        // LED 33
_delay_ms(5);
PORTA = 0b11111101;
PORTB = 0b11111111;
PORTD = 0b00000100;        // LED 34
_delay_ms(5);
PORTA = 0b11111011;
PORTB = 0b11111111;
PORTD = 0b00000100;        // LED 35
_delay_ms(5);
PORTA = 0b11110111;
PORTB = 0b11111111;
PORTD = 0b00000100;        // LED 36
_delay_ms(5);
PORTA = 0b11101111;
PORTB = 0b11111111;
PORTD = 0b00000100;        // LED 37
_delay_ms(5);
PORTA = 0b11011111;
PORTB = 0b11111111;
PORTD = 0b00000100;        // LED 38
_delay_ms(5);
PORTA = 0b10111111;
PORTB = 0b11111111;
PORTD = 0b00000100;        // LED 39
```

```c
_delay_ms(5);
PORTA = 0b01111111;
PORTB = 0b11111111;
PORTD = 0b00000100;        // LED 40
_delay_ms(5);
PORTA = 0b11111111;
PORTB = 0b11111110;
PORTD = 0b00000100;        // LED 41
_delay_ms(5);
PORTA = 0b11111111;
PORTB = 0b11111101;
PORTD = 0b00000100;        // LED 42
_delay_ms(5);
PORTA = 0b11111111;
PORTB = 0b11111011;
PORTD = 0b00000100;        // LED 43
_delay_ms(5);
PORTA = 0b11111111;
PORTB = 0b11110111;
PORTD = 0b00000100;        // LED 44
_delay_ms(5);
PORTA = 0b11111111;
PORTB = 0b11101111;
PORTD = 0b00000100;        // LED 45
_delay_ms(5);
PORTA = 0b11111111;
PORTB = 0b11011111;
PORTD = 0b00000100;        // LED 46
_delay_ms(5);
```

```
PORTA = 0b11111111;
PORTB = 0b10111111;
PORTD = 0b00000100;     // LED 47
_delay_ms(5);
PORTA = 0b11111111;
PORTB = 0b01111111;
PORTD = 0b00000100;     // LED 48
_delay_ms(5);
PORTA = 0b11111111;
PORTB = 0b11111111;
PORTD = 0b00000100;     // blank
_delay_ms(5);
PORTA = 0b11111110;
PORTB = 0b11111111;
PORTD = 0b00001000;     // LED 49
_delay_ms(5);
PORTA = 0b11111101;
PORTB = 0b11111111;
PORTD = 0b00001000;     // LED 50
_delay_ms(5);
PORTA = 0b11111011;
PORTB = 0b11111111;
PORTD = 0b00001000;     // LED 51
_delay_ms(5);
PORTA = 0b11110111;
PORTB = 0b11111111;
PORTD = 0b00001000;     // LED 52
_delay_ms(5);
PORTA = 0b11101111;
```

```c
PORTB = 0b11111111;
PORTD = 0b00001000;     // LED 53
_delay_ms(5);
PORTA = 0b11011111;
PORTB = 0b11111111;
PORTD = 0b00001000;     // LED 54
_delay_ms(5);
PORTA = 0b10111111;
PORTB = 0b11111111;
PORTD = 0b00001000;     // LED 55
_delay_ms(5);
PORTA = 0b01111111;
PORTB = 0b11111111;
PORTD = 0b00001000;     // LED 56
_delay_ms(5);
PORTA = 0b11111111;
PORTB = 0b11111110;
PORTD = 0b00001000;     // LED 57
_delay_ms(5);
PORTA = 0b11111111;
PORTB = 0b11111101;
PORTD = 0b00001000;     // LED 58
_delay_ms(5);
PORTA = 0b11111111;
PORTB = 0b11111011;
PORTD = 0b00001000;     // LED 59
_delay_ms(5);
PORTA = 0b11111111;
PORTB = 0b11110111;
```

```c
        PORTD = 0b00001000;        // LED 60
        _delay_ms(5);
        PORTA = 0b11111111;
        PORTB = 0b11101111;
        PORTD = 0b00001000;        // LED 61
        _delay_ms(5);
        PORTA = 0b11111111;
        PORTB = 0b11011111;
        PORTD = 0b00001000;        // LED 62
        _delay_ms(5);
        PORTA = 0b11111111;
        PORTB = 0b10111111;
        PORTD = 0b00001000;        // LED 63
        _delay_ms(5);
        PORTA = 0b11111111;
        PORTB = 0b01111111;
        PORTD = 0b00001000;        // LED 64



    }
}
```

Schematic Diagram

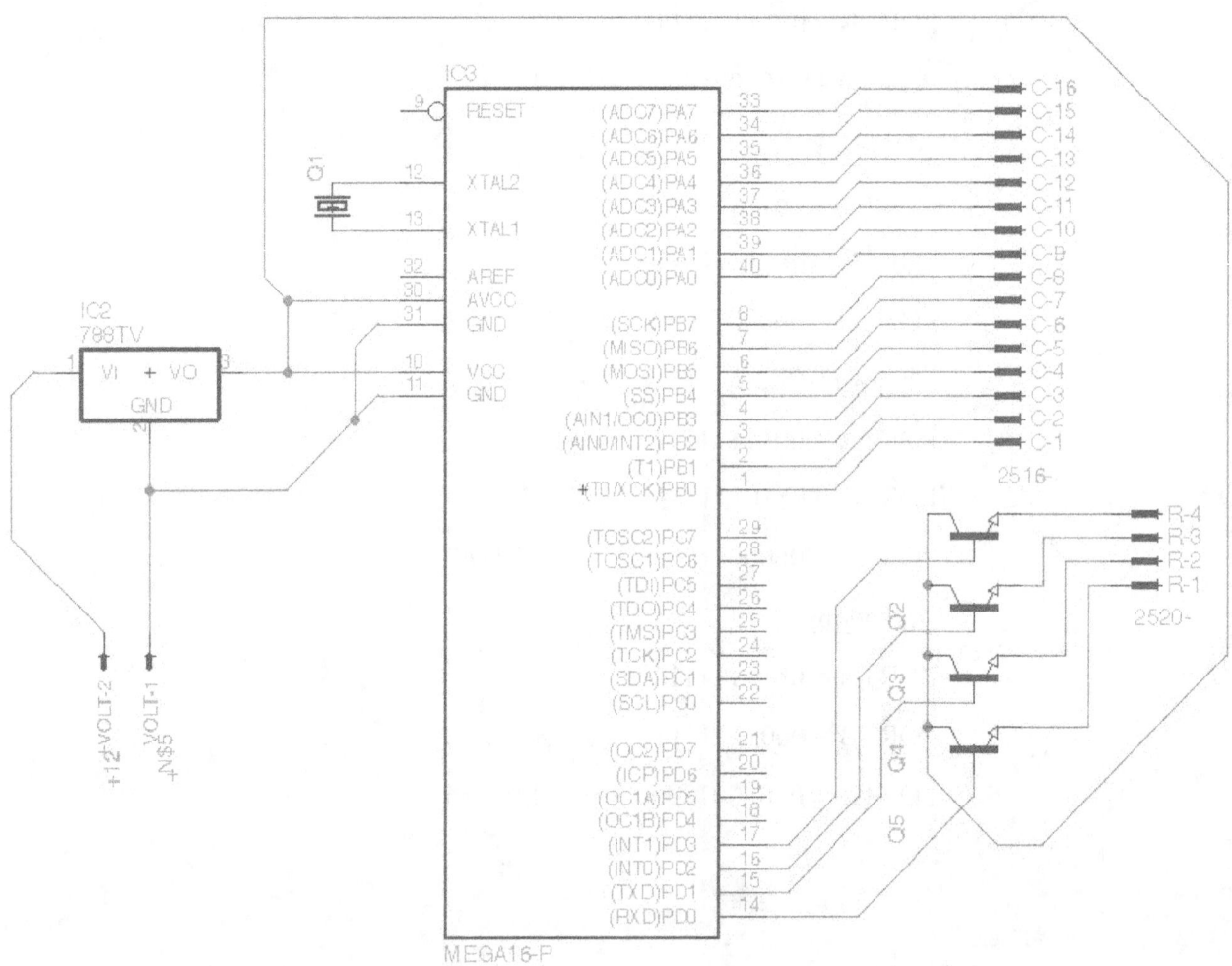

Experiment – 24

Sound Sensor Interfacing

In this practice we are going to introduce sound sensor it is a small board that combines a microphone and some processing circuitry. It provides not only an audio output, but also a binary indication of the presence of sound, and an analog representation of it's amplitude.

The Sound Detector is an analog circuit, and as such, it's more sensitive to noise on the power supply than most digital circuits. Since the capsule is effectively a voltage divider sitting across the power rails, it will transcribe any noise on the supply lines onto the capsule output. The next circuit in the chain is a high-gain amplifier, so any noise on the supply will then be amplified.

Therefore, the Sound Detector may require more careful power supply configuration than many circuits.

The sensor consist of microphone which is type of tranducer which converts the audio signal to electric signal, the sensitivity of sensor depend on pot which is variable resistor and can set the sensetiviy of signal. Sensor gives the output in Analog form hence can easily inter face with Atmega controller.

In testing with various supplies, a significant degree of variability was discovered - some supplies are less noisy than others. One exhibited as much as 30 mV ripple on the supply output, an as a result, the Sound Detector was rather sensitive and unstable. You can check how clean a power supply is by checking it with an oscilloscope or volt meter, set to the AC Volts (or, if provided, the AC millivolts) range. A truly clean supply will show 0.000 VAC. Based on the supplies used in testing, ripple of more than about 10 mV is problematic.

Amplitude Calibration

The Sound Detector comes set for moderate sensitivity - speaking directly into the microphone, or clapping your hands nearby should cause the gate output to fire. If you find that it doesn't work well in a specific application, you can change the circuit to be more or less sensitive.

The heart of the Sound Detector is the electret microphone capsule – without it, we couldn't convert acoustic energy into electrical energy. These capsules have a couple of quirks that we need to understand in order to apply them successfully.

Inside some of digital sensor the capsule is the diaphragm, which is actually one plate of a small capacitor. That capacitor forms a voltage divider with the external bias resistor. The diaphragm moves in response to sound, and the capacitance changes as the plates get closer together or farther apart, causing the divider to change. Since capacitors are sensitive to loading, it's internally buffered with a JFET (junction field-effect transistor).

Due to the mechanical and electronic tolerances involved, some capsules are more sensitive than others. Also, the JFET is rather sensitive to noise on the power supply. Both of these factors need to be accounted for when deploying the Sound Detector. Basically we recommend to make use of analog sensors.

Note:-

The command line written in blue color like – "lcd_clrscr();" are temporary program line to understand the analog value given out from sensor, once you calibrate the analog value for sensor note this value in program for condition such as at my end for forward direction x was triggering at greater than 410 so I have put it in condition "(x > 680)". When you wll calibrate all analog value **you can delete this line or can make comment to it**.

Component details

1) AVR ATmega16 IC3
2) 7805 IC2
3) Crystal 16 MHz
4) LCD 16 x 2
5) Sound Sensor
6) buzzer

7) 12 volt battery adaptor

8) ISP burner

Program

#include<avr/io.h>

#include<swits.h>

#include<util/delay.h>

main()

{

    int x;

    init_adc();

    lcd_init();

    while(1)

    {

        x=read_adc(0);

        _delay_ms(150);

        lcd_clrscr();

        lcd_goto(1,1);

        lcd_printi(x);

        _delay_ms(150);

        if(x > 680)

        {

        lcd_clrscr();

        lcd_goto(1,1);

        lcd_prints("Low sound");

```c
            _delay_ms(5000);
        }



else    if(x > 400 && x < 679)
        {
        lcd_clrscr();
        lcd_goto(1,1);
        lcd_prints("Medium sound");
        _delay_ms(5000);


        }
else    if(y < 399)
        {
        lcd_clrscr();
        lcd_goto(1,1);
        lcd_prints("High sound");
        _delay_ms(5000);


        }
else
        {
        lcd_clrscr();
        lcd_goto(1,1);
        lcd_prints("Sound sensor");
        _delay_ms(150);


        }
```

}
}

Schematic Diagram

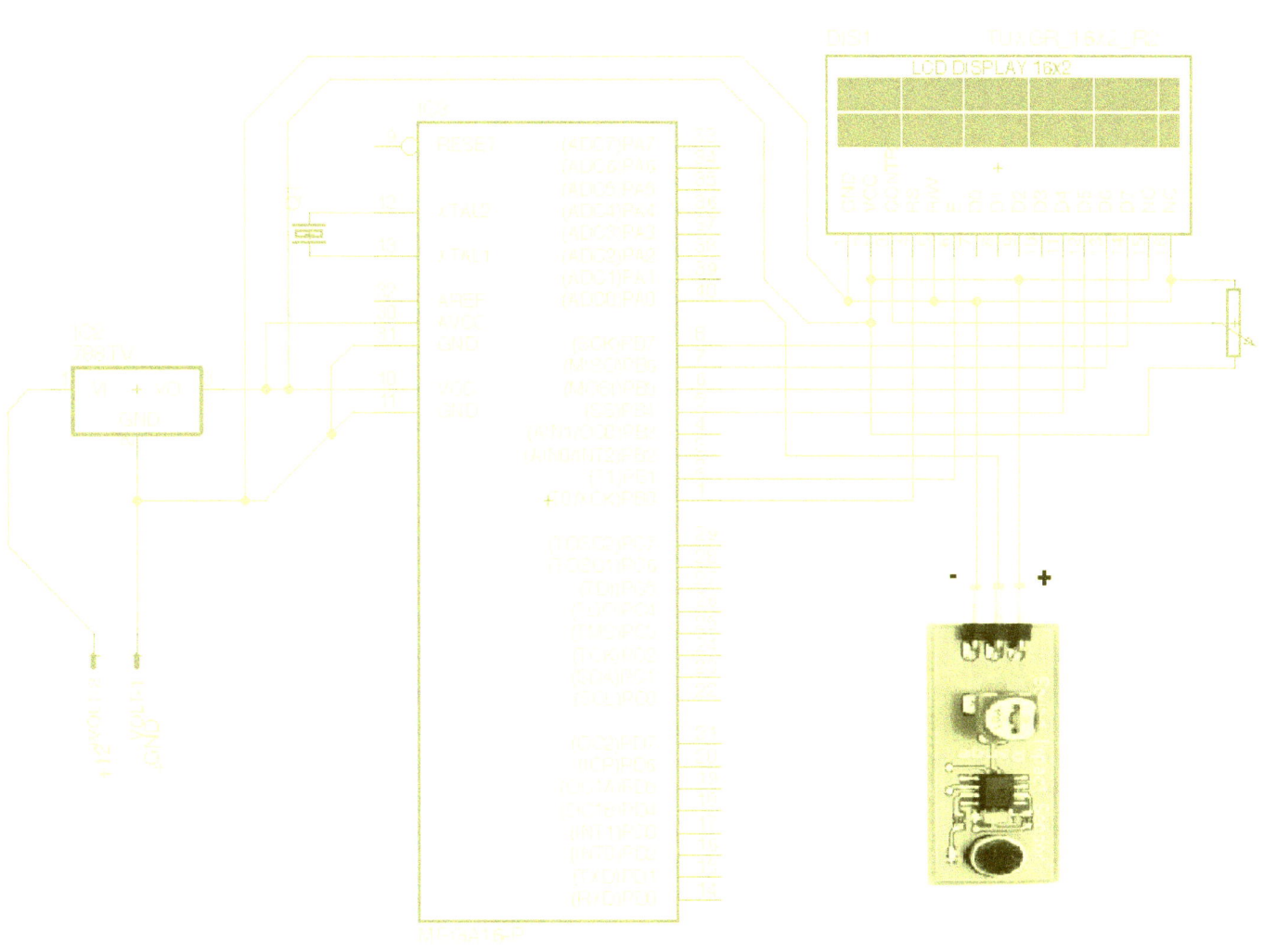

Experiment – 25

Accelerometer Interfacing

In this practice we are going to study Accelerometer. It is an electromechanical device that will measure acceleration forces. These forces may be static, like the constant force of gravity pulling at your feet, or they could be dynamic – caused by moving or vibrating the accelerometer. Accelerometers are of two types Analog and Digital. In this practice we will be discussing about Analog accelerometer. They give voltage as output which is proportional to acceleration. The digital one gives the PWM output or direct binary digital data

Sensor can measure static acceleration of gravity in tilt-sensing applications, as well as dynamic acceleration resulting from motion, shock, or vibration. The accelerometer sensor is used in mobile devices, gaming systems, disk drive protection, image stabilization, sports devices, health devices and etc. The analog outputs are ratio metric: that means that 0g measurement output is always at half of the 3.3V output (1.65V), -3g is at 0V and 3g is at 3.3V with full scaling in between. The accelerometer sensor gives three analog output values which correspond to three coordinate axes: – x-axis, y-axis and z-axis values.

we will practicing to interface a 3-axis accelerometer sensor with AVR ATmega16 microcontroller. Here, we will measure the tilt of the accelerometer sensor or the material to which the accelerometer sensor is attached and we will display the three outputs of accelerometer sensor in a 16X2 alphanumeric LCD. But, the three outputs of 3-axis accelerometer sensor are analog in nature and microcontroller cannot process the analog signal directly. So, first it will convert the three analog outputs of accelerometer sensor to digital values using its analog to digital converter and then it will display the three converted digital values in the 16X2 alphanumeric LCD. Now, we will tilt the accelerometer sensor or the material to which the sensor is attached in different direction and we will see the changes in its output values in the 16×2 alphanumeric LCD.

Note:-

The command line written in blue color like – "lcd_clrscr();" are temporary program line to understand the analog value given out from sensor, once you calibrate the analog value for sensor note this value in program for condition such as at my end for forward direction x was triggering at greater than 410 so I have put it in condition "(x > 410)". When you wll calibrate all analog value **you can delete this line or can make comment to it**.

Component details

1) AVR ATmega16 IC3

2) 7805 IC2

3) Crystal 16 MHz

4) LCD 16 x 2

5) Accelerometer

6) buzzer

7) 12 volt battery adaptor

8) ISP burner

Program

```c
#include<avr/io.h>
#include<swits.h>
#include<util/delay.h>
main()
{
    int x,y,z,w;
    init_adc();
    lcd_init();

    while(1)
    {
        x=read_adc(0); // read the ADC 0
        y=read_adc(1); // read the ADC 1
        z=read_adc(2); // read the ADC 2

        _delay_ms(150);
        lcd_clrscr();
        lcd_goto(1,1);
        lcd_printi(x);
        lcd_goto(1,2);
        lcd_printi(y);
        lcd_goto(8,1);
        lcd_printi(z);
```

```c
if(x > 410)              // condition for right
    {
    lcd_clrscr();
    lcd_goto(1,1);
    lcd_prints("right");
    }


else   if(x < 330)       // condition for left
    {
    lcd_clrscr();
    lcd_goto(1,1);
    lcd_prints("Left");

    }
else   if(y < 330)       // condition for reverse
    {
    lcd_clrscr();
    lcd_goto(1,1);
    lcd_prints("Reverse");

    }
else   if(y > 450)       // condition for forward
    {
    lcd_clrscr();
    lcd_goto(1,1);
    lcd_prints("forward");

    }
```

```
else    if(z < 310)     // condition for up
        {
        lcd_clrscr();
        lcd_goto(1,1);
        lcd_prints("Up");

        }
else    if(z > 390)     // condition for down
        {
        lcd_clrscr();
        lcd_goto(1,1);
        lcd_prints("Down");

        }
else
        {
        lcd_clrscr();
        lcd_goto(1,1);
        lcd_prints("Acclerometer");

        }

    }
}
```

Schematic Diagram

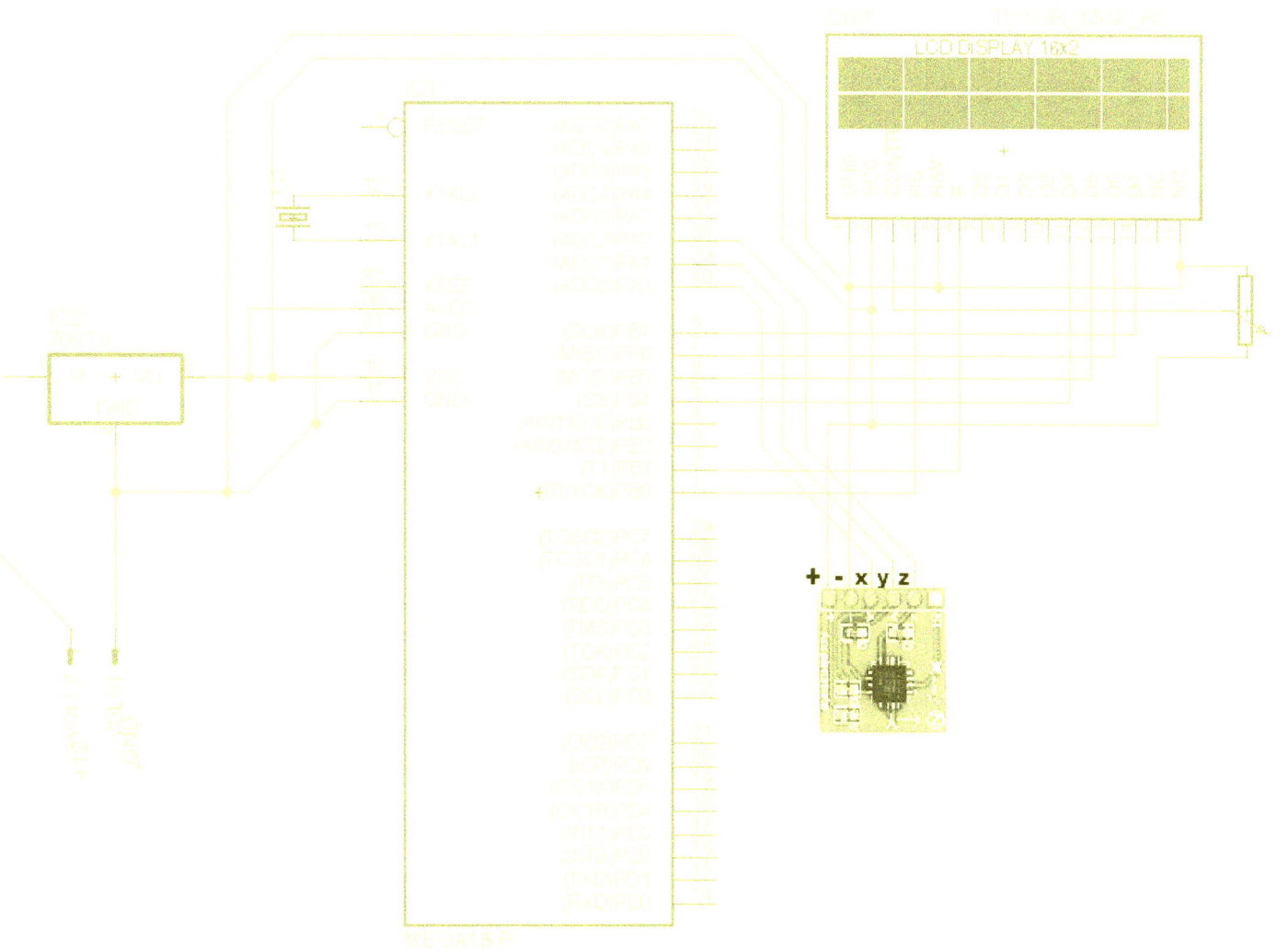

Appendix

Introduction to AVRstudio

1.1 Introduction

Welcome to AVR Studio from Atmel Corporation. AVR Studio is a Development Tool for the AT90S Series of AVR microcontrollers. This manual describes the how to install and use AVR Studio. AVR Studio enables the user to fully control execution of programs on the AT90S In-Circuit Emulator or on the built-in AVR Instruction Set Simulator. AVR Studio supports source level execution of Assembly programs assembled with the Atmel Corporation's AVR Assembler and C programs compiled with IAR Systems' ICCA90 C Compiler for the AVR microcontrollers. AVR Studio runs under Microsoft Windows95 and Microsoft Windows NT.

1.2 Installing

AVR Studio AVR Studio is delivered on two diskettes. Note that in some cases, the second diskette will not be asked for by the installation program. This is because some of the files required to run AVR Studio may already be present in the system. In order to install AVR Studio under Windows95 and Windows NT 4.0: 1. Insert the diskette labeled AVR Studio Diskette 1 in drive A: 2. Press the Start button on the Taskbar and select Run 3. Enter "A:SETUP" in the Open field and press the OK button 4. Follow the instructions in the Setup program In order to install AVR Studio under Windows NT 3.51: 1. Insert the diskette labeled AVR Studio Diskette 1 in drive A: 2. Select Run from the File menu 3. Enter "A:SETUP" in the Command Line field and press the OK button 4. Follow the instructions in the Setup program Once AVR Studio has been installed, it can be started by double clicking the AVR Studio icon. If an Emulator is the desired execution target, remember to connect the AVR InCircuit Emulator before starting AVR Studio.

1.3 Description

This section gives a brief description of the main features of AVR Studio. AVR Studio enables execution of AVR programs on an AVR In-Circuit Emulator or the built-in AVR Instruction Set Simulator. In order to execute a program using AVR Studio, it must first be compiled with IAR Systems' C Compiler or assembled with Atmel's AVR Assembler to generate an object file which can be read by AVR Studio. Rev. 1019A-A–01/98 An example of what AVR Studio may look like during execution of a program is shown below. In addition to the Source window, AVR Studio defines a number of other windows which can be used for inspecting the different resources on the microcontroller.

The key window in AVR Studio is the Source window. When an object file is opened, the Source window is automatically created. The Source window displays the code currently being executed on the execution target (i.e. the Emulator or the Simulator), and the text marker is always placed on the next statement to be executed. The Status bar indicates whether the execution target is the AVR In-Circuit Emulator or the built-in Instruction Set Simulator. By default, it is assumed that execution is done on source level, so if source information exists, the program will start up in source level mode. In addition to source level execution of both C and Assembly programs, AVR Studio can also view and execute programs on a disassembly level. The user can toggle between source and disassembly mode when execution of the program is stopped. All necessary execution commands are available in AVR Studio, both on source level and on disassembly level. The user can execute the program, single step through the code either by tracing into or stepping over functions, step out of functions, place the cursor on a statement and execute until that statement is reached, stop the execution, and reset the execution target. In addition, the user can have an unlimited number of code breakpoints, and every breakpoint can be defined as enabled or disabled. The breakpoints are remembered between sessions. The Source window gives information about the control flow of the program. In addition, AVR Studio offers a number of other windows which enables the user to have full control of the status of every element in the execution target. The available windows are: 1. Watch window: Displays the values of defined symbols. In the Watch window, the user can watch the values of for instance variables in a C program. 2. Register window: Displays the contents of the register file. The registers can be modified when the execution is stopped.

3. Memory windows: Displays the contents of the Program Memory, Data Memory, I/O Memory or EEPROM Memory. The memories can be viewed as hexadecimal values or as ASCII characters. The memory contents can be modified when the execution is stopped. 4. Peripheral windows: Displays the contents of the status registers associated with the different peripheral devices: • EEPROM Registers • I/O Ports • Timers • etc. 5. Message window: Displays messages from AVR Studio to the user 6. Processor window: Displays vital information about the execution target, including Program Counter, Stack Pointer, Status Register and Cycle Counter. These parameters can be modified when the execution is stopped. The first time an object file is being executed, the user needs to set up the windows which are convenient for observing the execution of the program, thereby tailoring the information on the screen to the specific project. The next time that object file is loaded, the setup is automatically reconstructed The different windows will be described more carefully in the next chapter.

1.4.1 Source window

The Source window is the main window in an AVR Studio session. It is created when an object file is opened, and is present throughout the session. If the Source window is closed, the session is terminated. The Source window displays the code which is being executed.

The next instruction to be executed is always marked by AVR Studio. If the marker is moved by the user, this next statement can still be identified since the previously marked text becomes red. A breakpoint is identified in the Source window as a dot to the left of the statement where the breakpoint is set. If the button to the right of the module selection box is pressed, the Source window switches between source level and disassembly level execution. When AVR Studio is in disassembly mode, all operations, such as Single stepping, is done on disassembly level. In some cases, no source level information is available, for instance if an Intel-Hex file is selected as the object file. When no source level information is available, execution must be done on disassembly level. The Toggle breakpoint, Run to Cursor and the Copy functions are also available by pressing the right mouse button in the Source window. When the right mouse button is pressed, a menu appears on the screen: If the cursor is placed on a statement and a Run to Cursor command is issued, the program will execute until it reaches the instruction where the cursor is placed. Breakpoints are set in a similar way: the cursor is placed on a statement, and a Toggle Breakpoint command is issued. If a breakpoint was already set on the statement, the breakpoint will be removed. If no breakpoint was set on the statement, a breakpoint is inserted. An object file can consist of several modules. Only one module is displayed at a time, but the user can change to the other modules by selecting the module of interest in the selection box on the top left of the Source window. This is a useful feature for viewing and setting breakpoints in other modules than the one currently active. The Source window supports the Windows Clipboard. The user can select parts of (or all) the contents in the Source window and then copy it to the Windows Clipboard by selecting Copy from the Edit menu..

1.4.2 Watch window

The Watch window can display the types and values of symbols like for instance variables in a C program. Since the AVR Assembler does not generate any symbol information, this window can only be used in a meaningful way when executing C programs. An example of a Watch window is given below. The Watch window has three fields. The first field is the name of the symbol which is being watched. The next is the type of the symbol, and the third is the value of the symbol. By default, the Watch window is empty, i.e. all the symbols the user would like to watch have to be added to the Watch window. Once a symbol has been added, it is remembered also in subsequent executions of the programs. The added watches are also remembered if the Watch window is closed. There are commands for adding watches, deleting watches and deleting all watches. A watch is added by giving an Add Watch command from the Watch menu or from the Debug toolbar. A watch can also be added by pressing the INS key if the Watch window is the active window. When an Add Watch command is issued, the user must enter the name of the symbol. The user can enter a symbol name with or without scope information. AVR Studio will first search for the symbol as if it contains scope information. If no such symbol is found, AVR Studio appends the symbol name to the current scope, and searches for this new symbol. If no such symbol is found, the symbol is unbound, "???" appears in the type field, and the value field remains empty. If the symbol name is found, the symbol is bound, the symbol with scope information is displayed in the watch field, and the type and value fields are filled out. Every time execution stops, AVR Studio tries to bind unbound symbols using the current scope. It is not possible to have floating symbols. Once a symbol is bound, it remains bound. The watches are remembered between sessions. Whether or not the symbol has been bound is a part of this information. If the program enters a scope where a bound symbol is not visible, the value field changes to "Out of scope". In order to delete a watch, the symbol name must first be clicked on using the left mouse button. When a symbol has been marked this way, AVR Studio accepts the Delete Watch command from the Watch menu. If the Watch window is the currently active window, the marked symbol can also be deleted by pressing the DEL key. The Watch window can be used for watching C arrays and structs as well as simple variables. The syntax is the same as in C (use braces ('[' and ']') for arrays and dot ('.') for structs). Dereferencing pointers is not supported. When watching arrays, variables can be used for dynamically indexing the arrays. It is for example possible to watch "my_array[i]" if i is an integer in the same scope as the array my_array. There can only be one Watch window active at a time. The watched symbols (with scope information) are remembered between sessions. The Watch window can also be toggled on and off, and the watches are also remembered if the Watch window is toggled on and off.

1.4.3 Register window

The Register window displays the contents of the 32 registers in the AVR register file When the Register window is resized, the contents is reorganized in order to best fit the shape of the window. The values in the Register window can be changed when the execution is stopped. In order to change the contents of a register, first make sure the execution is stopped. Then place the cursor on the register to change, press the left mouse button twice (not a double click, make sure to make a pause between the mouse button clicks). The register can then be changed. Type in the new contents in hexadecimal form. Finally, press the Enter key to confirm or the ESC key to cancel the change. Only one Register window can be active at a time.

1.4.4 Message window

The Message window displays messages from AVR Studio to the user. When a Reset command is issued, the contents of the Message window is cleared. An example of a Message window is given below. The contents in the Message window is remembered also when the Message window is toggled off and then on again. Only one Message window can be active at a time.

1.4.5 Memory window

The Memory window enables the user to inspect and modify the contents of the various memories present in the execution target. The same window is used to view all memory types. The Memory window can be used to view Data memory, Program memory, I/O memory and EEPROM memory. The user can have several concurrent Memory windows. An example of a Memory window is shown below. Which Memory type to view can be changed in the memory selection box at the top left of the Memory window. When a new Memory window is created, Data memory is the default memory type. AVR Studio not only keeps track over where the Memory windows are placed, but also which memory type it is displaying, and also the formatting status of the Window. A hexadecimal representation of the addresses and the contents of the memory is always displayed. In addition, the user can view the memory contents as ASCII characters. The user also has the option to group the hexadecimal representation into 16 bit groups in stead of 8 bit groups. When viewing Program memory, it is the Word address which is displayed in the address column, and the MSB is listed before the LSB in the data column.

1.4.5.1 Modifying memory

The user can modify the contents of the memories by issuing a double click on the line containing the item(s) to be changed. When a line in the Memory view is doubleclicked, a Window appears on the screen. If memory is viewed in 8 bit groups, the modifications are done on 8 bit groups and when memory is viewed as 16 bit groups, the modifications are done on 16 bit groups. When operating on 8 bit groups, the following Window appears: When operating on 16 bit groups, the following Window appears: The operation is the same in the two cases. If the Cancel button is pressed, no update is done even if the user has edited one or more of the values. If the OK button is pressed, the Memory is updated if one or more of the values are changed.

1.4.6 Processor window

The Processor window contains vital information about the execution target. An example of a Processor window is shown below. The Program Counter indicates the address of the next instruction to be executed. The Program Counter is displayed in hexadecimal form, and can be changed when the execution is stopped. When the Program Counter is changed, the current instruction is discarded. After the Program Counter is changed, the user must press the Single step function to jump to the desired address. The Stack Pointer holds the current value of the Stack Pointer which is placed in the I/O area. If the Target has a Hardware stack instead of an SRAM based stack, this is indicated in the Stack Pointer field. The Stack Pointer value can be changed when the execution is stopped. The Cycle Counter gives information about the number of clock cycles elapsed since last reset. The AVR In-Circuit Emulator does at time being not support a cycle counter so the Cycle Counter is always zero when the Emulator is the execution target. The Cycle Counter value is displayed as a decimal value and can be changed when the execution is stopped. The Flags is a display of the current value of the Status register. When the execution is stopped, these bits can be changed by clicking on the flags to change. A checked flag indicates that the flag is set (the corresponding bit in the Status register has the value 1). Only one Processor window can be active at a time.

1.4.7 Peripheral Device windows

The user can watch the contents of the I/O in the Memory window. Viewing the I/O area as a flat memory structure is not a very convenient way of observing the status of the many I/O devices of the microcontroller in question. Specialized Device windows have therefore been incorporated to ease the observation of I/O devices.

1.4.7.1 Timer/Counter 0

The Timer/Counter 0 window displays all essential information about Timer/Counter 0. When the Timer/Counter 0 is selected from the View → Peripherals → AT90SXXXX → Timer 0 menu, the following window appears on the screen: The Timer/Counter field gives the value of Timer/Counter 0. The prescaler field gives the value of the corresponding prescaler. The Overflow Flag check box and the Overflow Interrupt Enable check boxes gives the status and control bits of Timer 0.

1.4.7.2 Timer/Counter 1

The Timer/Counter 1 window displays all essential information about Timer/Counter 1. When the Timer/Counter 1 is selected from the View → Peripherals → AT90SXXXX → Timer 1 menu, the following window appears on the screen: The Timer/Counter 1 window displays in detail all the different parameters of Timer/Counter 1. A description of the different features of Timer/Counter 1 is given in the AVR Data Book. All the values can be changed when execution is stopped.

1.4.7.3 Port window

The Port window displays the three different I/O registers usually associated with a port. WHen the user selects a port from the View → Peripherals → AT90SXXXX → Port menu, the corresponding Port window appears: The Port window displays the setting of the Port, Pin and Data Direction Registers from the I/O area, both as hexadecimal values and as single bits. WHen execution is stopped, the values of the registers can be changed.

1.4.7.4 EEPROM Registers

When EEPROM Registers is selected from the View → Peripherals → AT90SXXXX → EEPROM Registers menu, the following window appears: AVR Studio knows how much EEPROM memory is available on the Target, so if required, the Address field contains the high byte of the address concatenated with the low byte of the address.

1.4.7.5 SPI window

The SPI window displays all essential information about the SPI. When the Timer/Counter 1 is selected from the View → Peripherals → AT90SXXXX → SPI menu, the SPI window appears on the screen: The SPI Data Register shows the SPI receive register. Editing the SPI Data Register value will not start sending data on the SPI even if the SPI is enabled. Initiating an SPI transfer can only be achieved by making the program running on the Target write to the SPI Data Register. The user can also observe and change the values of all the bits in the control and status registers.

1.4.7.6 UART window

The UART window displays all essential information about the UART. When the UART is selected from the View → Peripherals → AT90SXXXX → UART menu, the UART window appears on the screen: The UART window displays the UART Data Register, Baud Rate Register and the bits of the Control and Status Registers. Writing to the UART Data Register will not initiate a data transfer. The Data register must be written by the program executing on the Target.

1.4.7.7 Terminal I/O window

The Terminal I/O window enables simulation of Terminal I/O communication with a program executing in AVR Studio. When the Terminal I/O is selected from the View → Terminal I/O menu, the following window appears on the screen: The Terminal I/O window is only available when using the Simulator and when using the IAR C Compiler to generate code. The Compiler must be set up to generate the Debug format with Terminal I/O (default). The output from the program will the be sent to the Output part of the Terminal I/O window and the input to the program will be read from the Input part of the Terminal I/O window. When a character is read from the Input part, it is removed from the input view. The window has a fixed size.

1.4.7.8 Trace window

The Trace window keeps track of the history of the program currently being executed. When the Trace window is selected from the View → Trace menu, the Trace window appears on the screen: The Trace window is only available when using the Simulator. The column to the left defines the synchronization points to the code. If a synchronization character is doubleclicked, the marker in the source window is placed accordingly to mark the actual instruction. If the source window is in source mode, the source level synchronization points are indicated, whereas in disassembly mode, all instructions can be used as synchronization points. The Trace buffer is 32K cycles deep. 1.5 Commands AVR Studio incorporates a number of different commands. The commands can be given in various ways: through menu selections, toolbar buttons and by keyboard shortcuts. This section describes the available commands, and how they are invoked.

1.5.1 Administrative

1.5.1.1 Opening files

When Open is selected from the File menu, a file selection dialog appears on the screen (note that AVR assumes the file extension .OBJ, so by default, only files with this extension are listed). The user must then select the object file to execute. Currently, the AVR Studio supports the following formats: ■ IAR UBROF ■ AVR Object Files generated by the Atmel AVR Assembler ■ Intel-Hex AVR Studio automatically detects the format of the object file. The four most recently used files are also available under the File menu and can be selected for loading directly. When opening the file, AVR Studio looks for a file with the same filename as the file selected but with the extension AVD. This is a file AVR Studio generates when a file is closed, and it contains information about the project, including window placement. If the AVD project file is not found, only a Source window is created. The AVD file also contains information regarding breakpoints. Breakpoints defined in the previous session are reinserted unless the object file is newer than the project file. In the latter case, the breakpoints are discarded. If source level information is available, the program is executed until the first source statement is reached.

1.5.1.2 Closing files

When Close is selected from the File menu, all the windows in a session are closed. AVR Studio also writes a file in the same directory as the object file, containing project information. The file has the same name as the object file, but has the extension AVD.

1.5.1.3 Copying text

The user can mark text in the Source window and transfer this to the Windows Clipboard by selecting Copy from the Edit menu.

1.5.1.4 Downloading configuration (Emulator only)

When no files are loaded into AVR Studio, the File menu contains the option Download Configuration. If an Emulator is connected to the computer, the following message appears. As can be seen from the warning, downloading configurations should only be done if instructed so by Atmel. The files to be downloaded holds configuration files for the Emulator, and all other files will result in a non-functional Emulator.

1.5.2 Execution Control

Execution commands are used for controlling the execution of a program. All execution commands are available through menus, shortcuts and the Debug toolbar.

1.5.2.1 Go

The Go command in the Debug menu starts (or resumes) execution of the program. The program will be executed until it is stopped (user action) or a breakpoint is encountered. The Go command is only available when the execution is stopped. Shortcut: F5

1.5.2.2 Break

The Break command in the Debug menu stops the execution of the program. When the execution is stopped, all information in all windows are updated. The Break command is only available when a program is executing. Shortcut: CTRL-F5

1.5.2.3 Trace

Into The Trace Into command in the Debug menu executes one instruction. When AVR Studio is in source mode, one source level instruction is executed, and when in disassembly level, one assembly level instruction is executed. After the Trace Into is completed, all information in all windows are updated. Shortcut: F11

1.5.2.4 Step Over

The Step Over command in the Debug menu executes one instruction. If the instruction contains a function call/subroutine call, the function/subroutine is executed as well. If a user breakpoint is encountered during Step Over, execution is halted. After the Step Over is completed, all information in all windows are updated. Shortcut:

1.5.2.5 Step Out

The Step Out command in the Debug menu executes until the current function has completed. If a user breakpoint is encountered during Step Over, execution is halted. If a Step Out command is issued when the program is on the top level, the program will continue executing until it reaches a breakpoint or it is stopped by the user. After the Step Out command is completed, all information in all windows are updated. Shortcut: SHIFT+F11

1.5.2.6 Run to Cursor

The Run to Cursor command in the Debug menu executes until the program has reached the instruction indicated by the cursor in the Source window. If a user breakpoint is encountered during a Run to Cursor command, execution is not halted. If the instruction indicated by the cursor is never reached, the program executes until it is stopped by the user. After the Run to Cursor command is completed, all information in all windows are updated. Shortcut: F7

1.5.2.7 Reset

The Reset command performs a Reset of the execution target. If a program is executing when the command is issued, execution will be stopped. If the user is in source level mode, the program will, after the Reset is completed, execute until it reaches the first source statement. After the Reset is completed, all information in all windows are updated. Shortcut: SHIFT+F5

1.5.3 Watches

When executing at C source level, the Watch window can be used for watching symbols. When executing object files generated by the Atmel AVR Assembler, no symbol information is present so the Watch window can not be used for displaying any information.

1.5.3.1 Adding watches

In order to insert a new watch, the user must select Add Watch from the Watch window, or press the Add Watch button on the Debug toolbar. If the Watch window is not present when the Add Watch command is given, the Watch window is created, and already defined watches are reinserted (if any). If the Watch window is the active window, a new watch can also be added by pressing the INS key.

1.5.3.2 Deleting watches

The user can delete a watch by first marking the symbol to be deleted in the Watch window and then give a Delete Watch command from the Watch menu or from the Debug toolbar. Selecting a watch is done by moving the mouse pointer to the name of the watch and pressing the left mouse button. If the Watch window is the active window, a marked symbol can also be deleted by pressing the DEL key.

1.5.3.3 Deleting all watches

The Delete all watches command is available from the Watch menu. When this command is issued, all defined watches are removed from the Watch window.

1.5.4 Breakpoints

The user can set an unlimited number of code breakpoints. The breakpoints are remembered between sessions unless a new object file has been generated. If the object file is newer than the project file, the breakpoints are discarded. When a breakpoint is set on a location, the breakpoint is indicated by a dot on the left side of the instruction.

1.5.4.1 Toggle Breakpoint

The Toggle Breakpoint command toggles the breakpoint status for the instruction where the cursor is placed. Note that this function is only available when the source view is the active view.

1.5.4.2 Clear all breakpoints

This function clears all defined breakpoints, including breakpoints which have been disabled.

1.5.4.3 Show list

When Show list is selected, the following dialog appears on the screen: In the Breakpoints dialog, the user can inspect existing breakpoints, add a new breakpoint, remove a breakpoint or enable/disable breakpoints.

1.5.5 Up/Download Memories

The user can download data to the SRAM and the EEPROM data memories. If the Up/Download Memories is selected from the File menu after a file has been loaded into AVR Studio, the following window appears on the screen: The Up/Download Memories function reads and writes Intel-Hex files. Use the Browse button to select the file to read from/write to. Then select the desired function.

1.5.6 Toolbars

AVR Studio contains three different toolbars described below. The toolbars can be individually removed and/or reinserted if desired by unchecking/checking them in the View → Toolbars menu.

Installation Process

Go to the folder location where the AVR Studio 4 installation file is located. Double Click the file and a setup dialog box opens which is shown in the below picture.

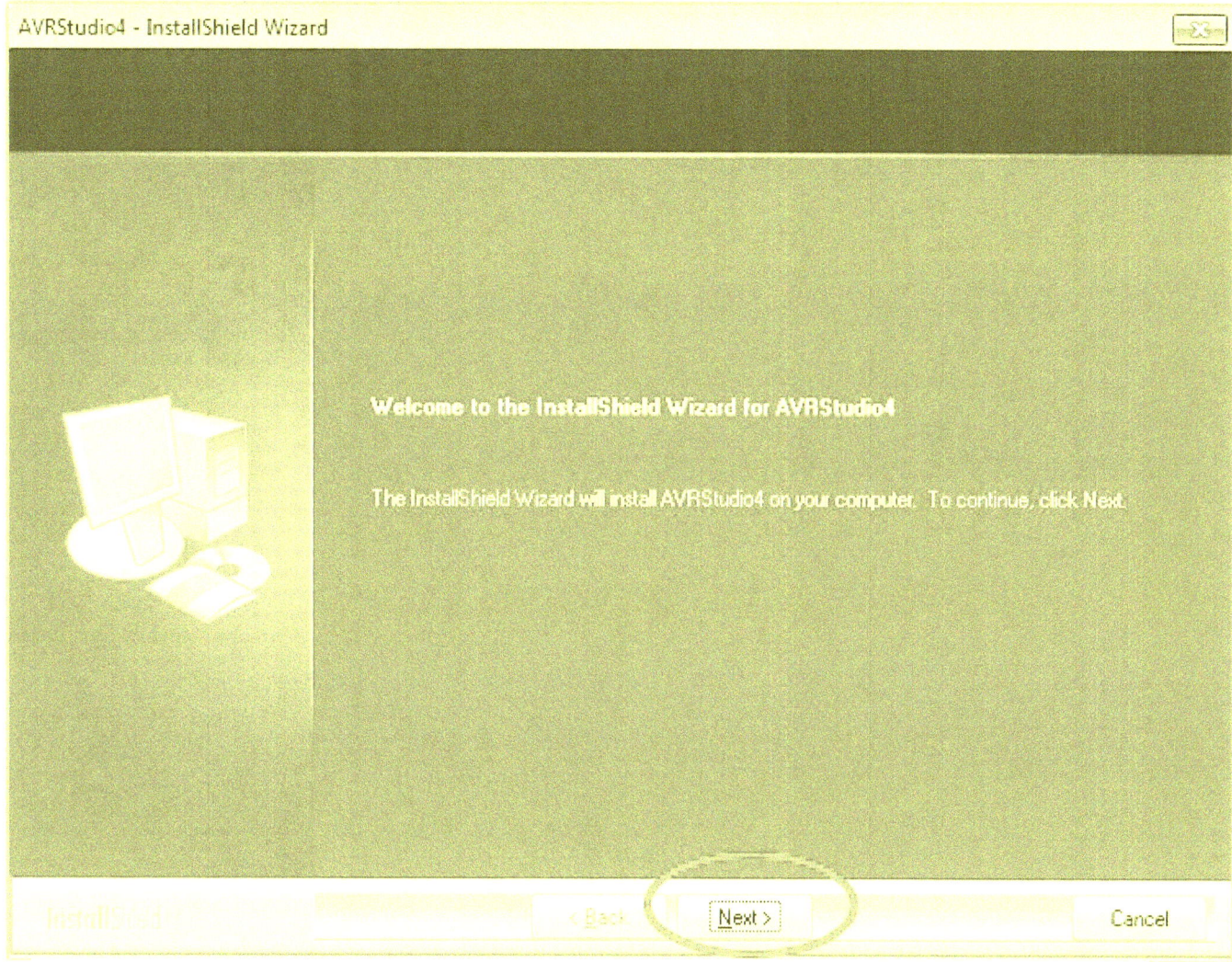

Click "Next" button to continue with the installation. The below picture shows the next dialog box:

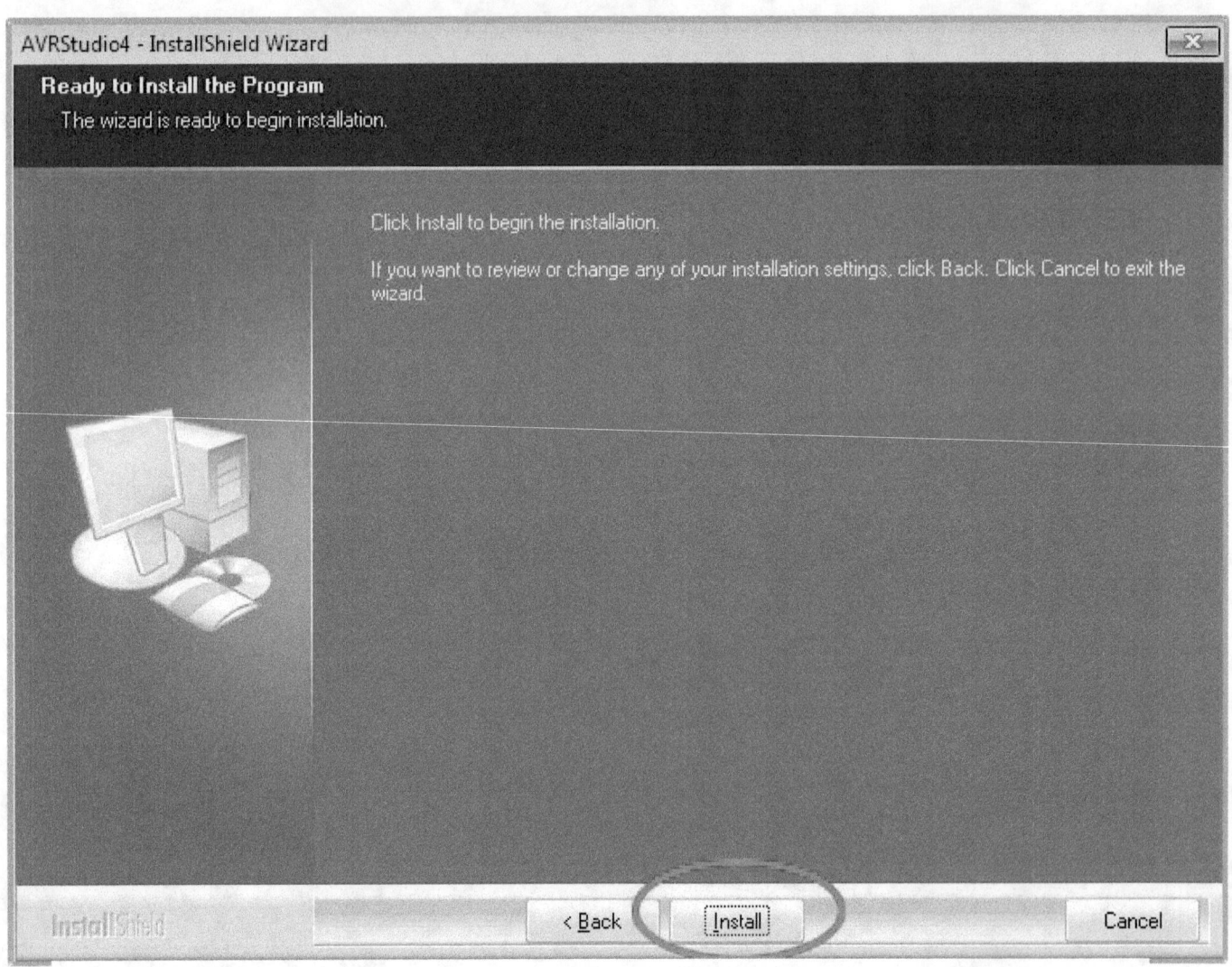

Check the "I accept the terms of the license agreement" radio button and Click "Next" button to continue with the Installation. The below picture shows the next dialog box:

Click "Next" button to continue with the installation. The below picture shows the next dialog box:

Check the "Install/upgrade Jungo USB Driver" check box to install Jungo USB driver and Click "Next" button to continue with the installation. The below picture shows the next dialog box:

Click "Install" button to start the final step of AVR Studio 4 installation. The below picture shows the next dialog box:

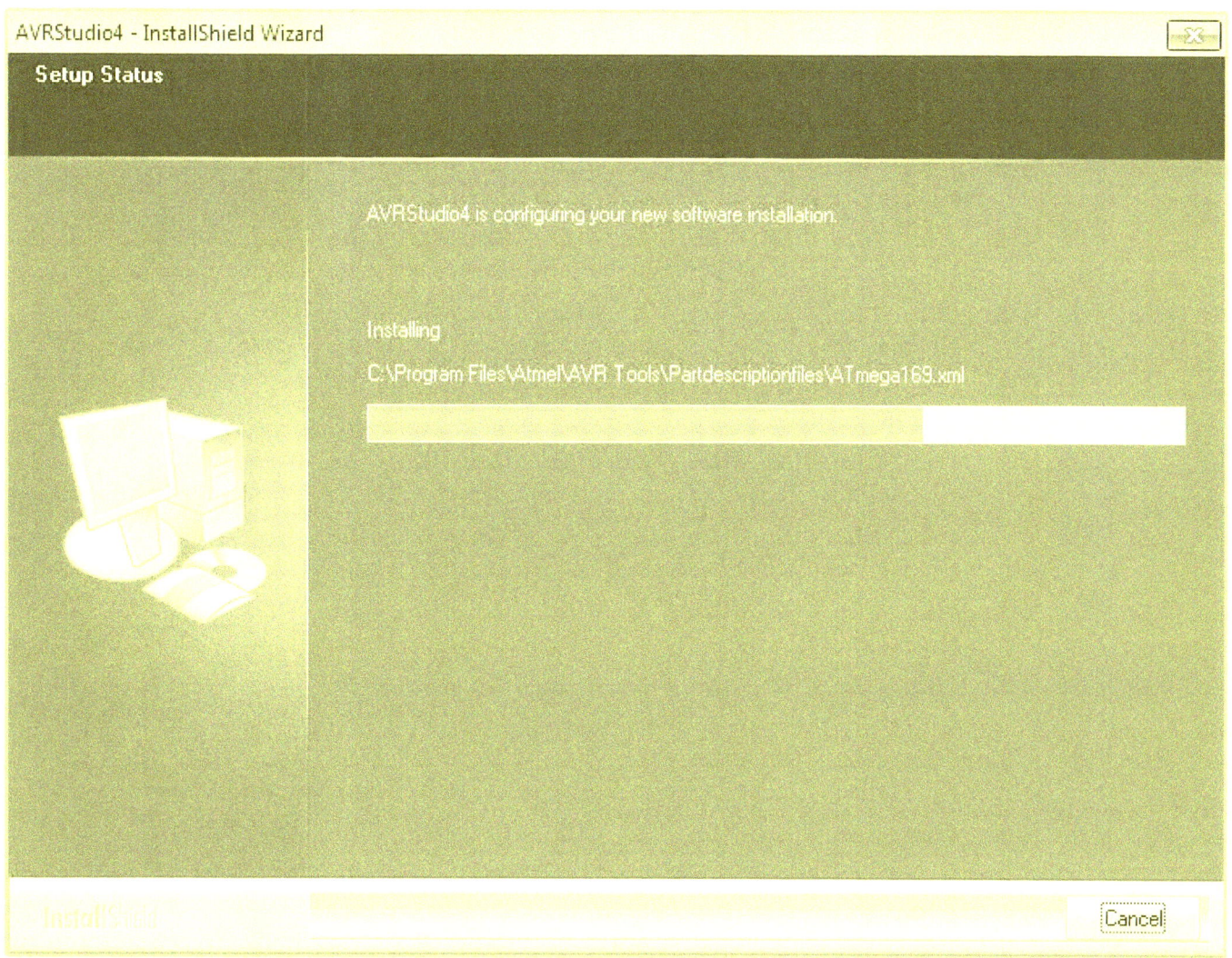

The Installation will take some minutes. Wait for the next dialog box to come up. The below pictures shows the dialog box when the installation completes.

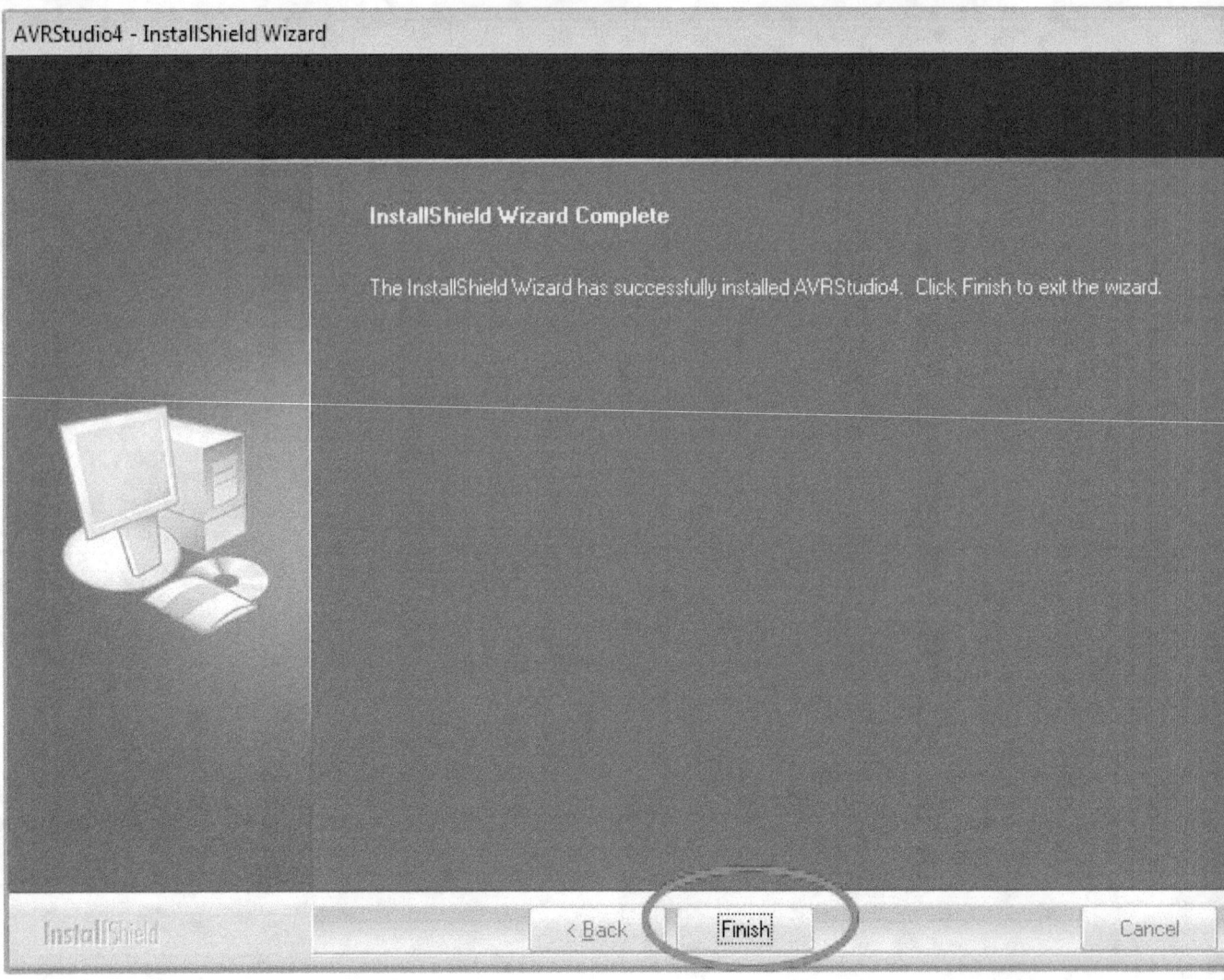

Click "Finish" button to complete the Installation of AVR Studio 4. Now, you have successfully completed the Installation of AVR Studio 4 and it is now ready for making AVR projects/applications

After installing AVRstudio we require to install Library which are get from "winAvr"

Installing WinAVR

WinAVR is a great GNU GCC cross-compiler suite that has all the necessary utilities to do AVR development in C (and assembly) language in Windows. "Cross-compiler" means that it runs on PC hardware, but generates binary code for another platform, in this case 8-bit AVR microcontrollers. GNU GCC is a widely spread, open-source C compiler (also C++ is often included) suite. WinAVR packs the compiler, several useful utilities such as make, and some AVR-specific stuff such as "avrdude" utility which can be used to flash the binary into AVR microcontroller, and change fuse settings of AVR MCUs.

Installing WinAVR is really straightforward: You just head to the WinAVR site, go to download section, and navigate yourself to the latest install.exe (at the time of writing, 2010-01-10 version) and run it. You can pretty much install it where you like, but for safety, one might want to avoid spaces in directory names, as they sometimes cause problams.
After you've installed WinAVR, restart command prompt for the PATH changes to take effect, and check that "avr-gcc", "avrdude" and "make" commands give their own error messages (avrdude actually just displays its command-line options) and you don't get the generic "command not recognized as…" rant. Now let's try our "make" process again:

Now if you had your USB tutorial project circuit attached to the computer via ISP programmer, and Makefile settings in order, you could just type "make flash" and start enjoying your program. If you don't stay tuned for next part which will cover compiling, linking and Makefile basics!

Burner to dump the Program

ISP Programmer

Introduction In-System Programming allows programming and reprogramming of any AVR microcontroller positioned inside the end system. Using a simple Three-wire SPI interface, the In-System Programmer communicates serially with the AVR microcontroller, reprogramming all non-volatile memories on the chip. In-System Programming eliminates the physical removal of chips from the system. This will save time, and money, both during development in the lab, and when updating the software or parameters in the field. This application note shows how to design the system to support In-System Programming. It also shows how a low-cost In-System Programmer can be made, that will allow the target AVR microcontroller to be programmed from any PC equipped with a regular 9-pin serial port. Alternatively, the entire In-System Programmer can be built into the system allowing it to reprogram itself. The Programming

Interface For In-System Programming, the programmer is connected to the target using as few wires as possible. To program any AVR microcontroller in any target system, a simple Six-wire interface is used to connect the programmer to the target PCB. The connections needed.

The Serial Peripheral Interface (SPI) consists of three wires: Serial ClocK (SCK), Master In – Slave Out (MISO) and Master Out – Slave In (MOSI). When programming the AVR, the In-System Programmer always operate as the Master, and the target system always operate as the Slave. The In-System Programmer (Master) provides the clock for the communication on the SCK Line. Each pulse on the SCK Line transfers one bit from the Programmer (Master) to the Target (Slave) on the Master Out – Slave In (MOSI) line. Simultaneously, each pulse on the SCK Line transfers one bit from the target (Slave) to the Programmer (Master) on the Master In – Slave Out (MISO) line. 8-bit RISC Microcontroller Application Note Rev. 0943E–AVR–08/08 2 0943E–AVR–08/08 AVR910. Six-wire Connection Between Programmer and Target System

To assure proper communication on the three SPI lines, it is necessary to connect ground on the programmer to ground on the target (GND). To enter and stay in Serial Programming mode, the AVR microcontroller reset line has to be kept active (low). Also, to perform a Chip Erase, the Reset has to be pulsed to end the Chip Erase cycle. To ease the programming task, it is preferred to let the programmer take control of the target microcontroller reset line to automate this process using a fourth control line (Reset). To allow programming of targets running at any allowed voltage (2.7 - 6.0 V), the programmer can draw power from the target system (VCC). This eliminate the need for a separate power supply for the programmer. Alternatively, the target system can be supplied from the programmer at programming time, eliminating the need to power the target system through its regular power connector for the duration of the programming cycle shows the connector used by this In-System Programmer to connect to the target system. The standard connector supplied is a 2 x 3 pin header contact, with pin spacing of 100 mils.. Recommended In-System Programming Interface Connector Layout (Top View) Hardware Design Considerations To allow In-System Programming of the AVR microcontroller, the In-System Programmer must be able to override the pin functionality during programming. This section describes the details of each pin used for the programming operation. GND The In-System Programmer and target system need to operate with the same reference voltage. This is done by connecting ground of the target to ground of the programmer. No special considerations apply to this pin. RESET The target AVR microcontroller will enter Serial Programming mode only when its reset line is active (low). When erasing the chip, the reset line has to be toggled to end the erase cycle. To simplify this operation, it is recommended that the target reset can be controlled by the In-System Programmer. Immediately after Reset has gone active, the In-System Programmer will start to communicate on the three dedicated SPI wires SCK, MISO, and MOSI. To avoid driver contention, a series resistor should be placed on each of the three dedicated lines if there is a possibility that external circuitry could be driving these lines. The value of the resistors should be chosen depending on the circuitry connected to the SPI bus.

Software to burn Program

To dump the program we required software there are ample of software, Extreme burner is simplified 1.

Download the software "Extreme burner AVR" from trusted web, and follow the instruction.

Step 1

Connect the USB Programmer to your PCs USB port. Make sure you connect it to that USB port in you installed it during its installation. Wait for a "ding" sound from PC. Now the programmer is installed correctly. The GREEN LED will glow to show programmer is ready.

Note

If windows says "USB Device not recognized" make sure the USB cables are not broken.

If windows says "New hardware found" you have connected the programmer to a different port than which you have installed. Or you have not yet installed the programmer ! Please see installation instruction in this CD.

Please disconnect the Programmer after programming to ensure that your computer is safe.

Step 2

Launch eXtreme Burner – AVR from Desktop Icon or Start Menu. You will get a screen similar to this.

From *Chip* Menu Select the MCU in use, say ATmega32

Select *File->Open Flash* or *Open* from Toolbar

Select the HEX file

EEPROM: If you want to program the on-chip EEPROM load a .eep file by selecting *File->Open EEPROM File*

Now turn on the target. Now every thing is set,
select *Write->All* or select *Write All* from toolbar to burn the device. While burning the RED LED will glow indicating BUSY state. If everything is setup properly, you will get the following message.

Header Files

Swits.h

Please copy the header file to C:\WinAVR-20100110\avr\include

```c
#include <avr/io.h>              // The necessary header file required for avr atmega 16/32
#include <util/delay.h>          // Header file required for Delay
#include <compat/deprecated.h>   // Used for the commands like sbi,cbi,bit_is_clear etc..
#include <stdlib.h>

#ifndef multiutil_H
#define multiutil_H

#define LCD_PORT PORTB           // PortA as LCD command port
#define LCD_DDR  DDRB

#define rs 0                     // connect RS pin to LCD_C_PORT ie PA0
                                 // connect RW pin to LCD_C_PORT ie PA6
#define en 1                     // connect EN pin to LCD_C_PORT ie PA1
```

```c
void lcd_init(void);                            // Function to init lcd ie to start display ,4 bit mode, cursor ON, etc

void write_command(unsigned char comm);         // Function to write command in lcd

void write_data(unsigned char lcd_data);        // Fuction to write data in lcd

void ilabs_lcd_prints(char *print_str);         // Function to print String On LCD

void lcd_printi(int num);

void lcd_clrscr(void);

void lcd_goto(uint8_t,uint8_t);

void lcd_printc(char);

void init_usart(void);
void transmit_char(char);
void transmit_string(char *);

void init_pwm(int mode);

void init_adc(void);
int read_adc(char adc);

void start_i2c(void);
void stop_i2c(void);
void init_i2c(void);
void write_i2c(char);
```

```c
char read_i2c(char);

#endif

//LCD
void lcd_init(void)
{
    _delay_ms(300);

    LCD_DDR=0b11110011;                   //Initialize the LCD port first

    write_command(0x80);                  //initial address of DDRAM for first line

    write_command(0x28);                  //To select 4 bit data

    write_command(0x0e);                  //Command to on the display & cursor blink/blink off

    write_command(0x01);                  //Command to clear LCD display

    write_command(0x06);                  //Command for character entry mode
}


void write_command(unsigned char comm)
{
    _delay_ms(2);
    LCD_PORT=((comm & 0xF0)|(1<<en));     //Sending 4 MSB bits of command, & Enable=1,RW=0,RS=0;
```

```c
        cbi(LCD_PORT,en);


        LCD_PORT=((comm<<4)|(1<<en));           //Sending 4 LSB bits of command
        cbi(LCD_PORT,en);
}



void write_data(unsigned char lcd_data)
{
        _delay_ms(2);

        LCD_PORT=((lcd_data & 0xF0)|(1<<en)|(1<<rs));  //Sending 4 MSB bits of command, & Enable=1,RW=0,RS=1;
        cbi(LCD_PORT,en);

        LCD_PORT=((lcd_data<<4)|(1<<en)|(1<<rs));       //Sending 4 LSB bits of command
        cbi(LCD_PORT,en);
}

void lcd_prints(char *print_str)
{
        while(*print_str)
        {
                if(*print_str=='\0')
                break;
                write_data(*print_str);
                print_str++;
        }
}
```

```c
void lcd_printi(int num)
{
        char buff[]={'0','0','0','0','0'};
        itoa(num,buff,10);
        lcd_prints(buff);
}


void lcd_printc(char data)
{
                if(data=='\n')
                        lcd_goto(0,2);
                write_data(data);
}
void lcd_goto(uint8_t x,uint8_t y)
{
        if(y==1)
                write_command(0x80|x);
        else if(y==2)
                write_command(0xC0|x);
}


void lcd_clrscr(void)
{
        write_command(0x01);
        _delay_ms(3);
        write_command(0x80);
}
```

```c
//USART
void init_usart()
{
    //Communication protocol set
    //with 1 stop bit & 8 data bits
    UCSRC=(1<<UCSZ1)|(1<<UCSZ0)|(1<<URSEL);

    //Reception & transmission is enabled
    //Reception of data will occur on interrupt
    UCSRB=(1<<RXEN)|(1<<TXEN)|(1<<RXCIE);

    //Baud rate = 9600
    UBRRH=0x00;
    UBRRL=103;
}

void transmit_char(char data)
{
    loop_until_bit_is_set(UCSRA,UDRE);
    UDR=data;
}

void transmit_string(char *data)
{
    while(*data)
    {
        transmit_char(*data);
        data++;
    }
```

}

```c
/* //PWM
void init_pwm(int mode)
{
    DDRD|=(1<<PD4)|(1<<PD5);
    TCCR1A|=(1<<COM1A1)|(1<<COM1B1);
    TCCR1B|=(1<<CS10)|(1<<CS11);
    switch(mode)
    {
        case 8:TCCR1A|=(1<<WGM10); TCCR1B|=(1<<WGM12);break;
        case 9:TCCR1A|=(1<<WGM11); TCCR1B|=(1<<WGM12);break;
        case 10:TCCR1A|=(1<<WGM10)|(1<<WGM11); TCCR1B|=(1<<WGM12);break;
        default:TCCR1A|=(1<<WGM10); TCCR1B|=(1<<WGM12);break;
    }
}
*/

//ADC
void init_adc(void)
{
    ADMUX=(1<<REFS0);
    ADCSRA=(1<<ADPS0)|(1<<ADPS1)|(1<<ADPS2)|(1<<ADEN);
}

int read_adc(char adc)
{
```

```c
        ADMUX&=(0b11100000);
        ADMUX|=(adc);
        //  ADMUX=0b01

        ADCSRA|=(1<<ADEN)|(1<<ADSC);
        while((ADCSRA&(1<<ADIF))==0);
        return ADC;
}


//I2C
void init_i2c(void)
{
        PORTC |= _BV(0)|_BV(1);
        TWBR=0x47;
}


void start_i2c(void)
{
        TWCR|=(1<<TWINT)|(1<<TWSTA)|(1<<TWEN);
        while((TWCR&(1<<TWINT))==0);
}


void stop_i2c(void)
{
        TWCR|=(1<<TWINT)|(1<<TWSTO)|(1<<TWEN);
}


void write_i2c(char data)
{
```

```c
        TWDR=data;
        TWCR=(1<<TWINT)|(1<<TWEN);
        while((TWCR&(1<<TWINT))==0);
}
char read_i2c(char last)
{
        if(last==0)
                TWCR=(1<<TWINT)|(1<<TWEN)|(1<<TWEA);
        else
        {
                TWCR=(1<<TWINT)|(1<<TWEN);
        }
        while((TWCR&(1<<TWINT))==0);
        return TWDR;
}
```

www.ingramcontent.com/pod-product-compliance
Lightning Source LLC
Chambersburg PA
CBHW080536190526
45169CB00012B/2155